Current Topics in Microbiology 165 and Immunology

Neuronal Growth Factors

Edited by M. Bothwell

With 14 Figures

Springer-Verlag
Berlin Heidelberg New York London
Paris Tokyo Hong Kong Barcelona

Professor Mark Bothwell, MD
Department of Physiology and Biophysics
University of Washington
Seattle, WA 98195
USA

ISBN 3-540-52654-4 Springer-Verlag Berlin Heidelberg New York
ISBN 0-387-52654-4 Springer-Verlag New York Berlin Heidelberg

Library of Congress Catalog Card Number 15-12910
Printed in Germany

Typesetting: Thomson Press (India) Ltd, New Delhi;
Offsetprinting: Saladruck, Berlin; Bookbinding: B. Helm Berlin.
2123/3020-543210 – Printed on acid-free paper

Preface

A host of environmental factors regulate the embryonic development of neurons. In adults, the survival of neurons, the regeneration of damaged axons, and plastic changes in axonal arborization are also controlled by a complex array of environmental cues. An important category of regulatory influences involves target-derived hormone-like peptides acting on neuronal cell surface receptors. Nerve growth factor (NGF) was the first such factor to be characterized (LEVI-MONTALCINI and HAMBURGER 1953) and has served as the model against which all similar factors are compared. A number of factors with properties similar to NGF have been described with activities toward differing populations of neurons. Many of these factors have been characterized only poorly at the biochemical level. However, several factors have been characterized to the extent that molecular clones are available and complete amino acid sequences are known. These include: three structurally related factors, NGF itself (SCOTT et al. 1983), brain-derived neurotrophic factor (BDNF) (LEIBROCK et al. 1989), and neurotrophin-3 (NT-3) (MAISONPIERRE et al. 1990; HOHN et al. 1990); ciliary neurotrophic factor (CNTF) (LIN et al. 1989) and a second set of structural homologs; acidic and basic fibroblast growth factors (aFGF, bFGF) (ABRAHAM et al. 1986; JAYE et al. 1986). Investigators have cloned the receptors for NGF (JOHNSON et al. 1986; RADEKE et al. 1987) and FGF (LEE et al. 1989).

As the NGF and FGF systems have been the most extensively characterized, the biology of these two peptide factors, or, more precisely, families of peptide factors, will be described in detail in this volume. It is our expectation that the information obtained from the NGF and FGF systems will be invaluable for understanding the functional properties of the numerous similar factors which remain to be characterized. The NGF and FGF systems also serve to illustrate an important point: NGF was originally discovered as a neurotrophic factor. However, as will be discussed in the following pages, it is becoming clear that NGF and the NGF receptor are involved in regulation of a variety of non-neural cell types. Conversely, while aFGF and bFGF were originally discovered as mitogens for non-neural cell types, it is now clear that they have potent neurotrophic activity. Epidermal growth factor is another mitogenic growth factor which has been reported to have neurotrophic activity (MORRISON et al. 1988). It is likely that many of the growing list of mitogenic growth factors will have specific regulatory functions for specific populations of post-mitotic neurons.

Mark Bothwell

References

Abraham JA, Whang JL, Tumolo A, Mergia A, Friedman J, Gospodarowicz D, Fiddes JC (1986) Human basic fibroblast growth factor: nucleotide sequence and genomic organization. EMBO J 5: 2523–2528

Hohn A, Leibrock J, Bailey K, Barde Y-A (1990) Identification and characterization of a novel member of the nerve growth factor/brain-derived neurotrophic factor family. Nature 344: 339–341

Jaye M, Howk R, Burgess W, Ricca G, Chiu I, Ravera M, O'Brien S, Modi W, Maciag T, Drohan WN (1986) Human endothelial cell growth factor: cloning, nucleotide sequence and chromosome localization. Science 233: 541–545

Johnson D, Lanahan A, Buck CR, Sehgal A, Morgan C, Mercer E, Bothwell M, Chao M (1986) Expression and structure of the human NGF receptor. Cell 47: 545–554

Lee PL, Johnson DE, Cousens LS, Fried VA, Williams LT (1989) Purification and complementary DNA cloning of a receptor for basic fibroblast growth factor. Science 245: 57–60

Leibrock J, Lottspeich F, Hohn A, Hofer M, Hengerer B, Masiakowski P, Thoenen H, Barde Y-A (1989) Molecular cloning and expression of brain-derived neurotrophic factor. Nature 341: 149–154

Levi-Montalcini R, Hamburger V (1953) A diffusable agent of mouse sarcoma producing hyperplasia of sympathetic ganglia and hyperneurotization of viscera in the chick embryo. J Exp Zool 123: 233–278

Lin L-F, Mismer D, Lile JD, Armes LG, Butler ET, Vannice JL, Collins F (1989) Purification, cloning and expression of ciliary neurotrophic factor (CNTF). Science 246: 1023–1025

Maisonpierre PC, Belluscio L, Squinto S, Ip NY, Furth ME, Lindsay RM, Yancopoulos GD (1990) Neurotrophin-3: a neurotrophic factor related to NGF and BDNF. Science 247: 1446–1451

Morrison RS, Keating RF, Moskal JR (1988) Basic fibroblast growth factor and epidermal growth factor exert differential trophic effects on CNS neurons. J Neurosci Res 21: 71–79

Radeke MJ, Misko TP, Hsu C, Herzenberg LA, Shooter EM (1987) Gene transfer and molecular cloning of the rat nerve growth factor receptor. Nature 325: 593–597

Scott J, Selby M, Urdea M, Quiroga M, Bell GI, Rutter W (1983) Isolation and nucleotide sequence of a cDNA of mouse nerve growth factor. Nature 302: 538–540

Table of Contents

List of Contributors

(You will find the addresses at the beginning of the respective contribution)

Structure and Biosynthesis of Nerve Growth Factor

M. FAHNESTOCK

1 Introduction

The first description of nerve growth factor (NGF) by LEVI-MONTALCINI and HAMBURGER (1953) was of a factor from mouse sarcoma cells (S-180 and S-37) that elicited extensive growth of chick embryo neurons. Shortly thereafter, a

Molecular Biology Department SRI International, 333 Ravenswood Avenue, Menlo Park, CA 94025, USA

Current Topics in Microbiology and Immunology, Vol. 165

serendipitous use of snake venom in the purification of this factor led COHEN to the discovery that NGF was present in moccasin snake venom as well (COHEN and LEVI-MONTALCINI 1956). The finding that the mouse submaxillary gland—a homologue of snake venom glands—stores even larger quantities of NGF than does the snake venom gland (COHEN 1960) led to the extensive analyses of mouse salivary NGF that continue today.

The ability to elicit nerve fiber outgrowth from cultured chick embryo sensory or sympathetic ganglia is still the most reliable bioassay for NGF activity. Despite a large number of tissues shown to exhibit this activity, there are few sources of mammalian NGF sufficiently rich for purification and molecular experimentation. The submaxillary gland of the male mouse has been the traditional source; the ability to purify significant amounts of this protein led to the development of antibodies for immunoassay (SUDA et al. 1978; KORSCHING and THOENEN 1983; LARKFORS and EBENDAL 1987) and to the cloning of the gene for murine NGF (SCOTT et al. 1983; ULLRICH et al. 1983). However, the limited quantities of NGF present in other species and tissues as well as the poor immunological cross-reactivity among NGFs from different species have hampered experimentation. In recent years, comparative study of NGF has intensified due to the highly conserved nucleic acid sequences of NGF among species. This review will focus on the comparative aspects of the structure, biosynthesis, and processing of NGF from a number of sources.

2 Structure of NGF

2.1 Murine NGF

The structure, synthesis, and biological activity of NGF have been extensively studied in the mouse, but little is known about its processing or regulation. High molecular weight NGF isolated from mouse submaxillary glands has a sedimentation coefficient of 7S, from which a molecular weight of 140 000 was calculated (VARON et al. 1968). The complex consists of three types of subunits (SERVER and SHOOTER 1977): β, γ, and α.

2.1.1 The β Subunit

The most extensively studied subunit, β-NGF, is responsible for the nerve growth-promoting activity. cDNA clones for the β subunit have been obtained by two research groups (ULLRICH et al. 1983; SCOTT et al. 1983), and the genomic sequence including promoter also has been determined (SELBY et al 1987b; ZHENG and HEINRICH 1988). β-NGF is present in the 7S complex as a noncovalently linked dimer of molecular weight 26 000; strong denaturing agents separate β-NGF into two identical chains 118 amino acids long (VARON and SHOOTER 1970). Both monomers and dimers are biologically active (FRAZIER et al. 1973a; STACH and SHOOTER 1974). Each chain contains three intrachain disulfide bonds, and reduction of these bonds causes complete loss of biological activity (GREENE and SHOOTER 1980; THOENEN and BARDE 1980). Although tyrosine residues are not required for activity, tryptophan residues are important for biological activity, since oxidation of Trp-21 causes loss of NGF activity and receptor binding (FRAZIER et al. 1973b; COHEN et al. 1980). A potential glycosylation site exists in the mature NGF protein at Asn-45. Although the molecule was thought not to be glycosylated (ANGELETTI et al. 1973), a glycosylated form exists in the mouse submaxillary gland in small amounts (MURPHY et al. 1989). The biological activity of the glycosylated form does not differ from the nonglycosylated form. The isoelectric point of the β subunit is 9.4, making β-NGF the most basic of the three subunits.

2.1.2 The γ Subunit

The two γ subunits of the 7S NGF complex are responsible for its proteolytic activity (GREENE et al. 1969). Both the amino acid and nucleotide sequences of the γ subunit have been determined, and they demonstrate that it belongs to the kallikrein family of trypsin-like serine proteases (THOMAS et al. 1981a; ULLRICH et al. 1984; EVANS and RICHARDS 1985). Like other kallikreins, the γ subunit has a molecular weight of 26000 and is glycosylated.

The γ subunit displays a high degree of electrophoretic heterogeneity; up to six different forms have been identified (STACH et al. 1976; THOMAS et al. 1981b). The different forms of the γ subunit contain either two or three chains connected by disulfide bridges. The two and three chain forms arise by endoproteolytic cleavage of a single chain protein at two sites. C-terminal basic residues located at these internal cleavage sites are removed from some molecules to create further heterogeneity. This heterogeneity is not due to the variability of the tissue source, however, since each submaxillary gland contains all six forms (SMITH et al. 1968). The esterase activities of the different forms of the γ subunit are all equivalent (SERVER and SHOOTER 1977). Although no biological significance has been demonstrated for the existence of six different forms, a 7S complex formed with γ^3 (a three chain form) is the most stable (THOMAS et al. 1981b; NICHOLS and SHOOTER 1983). Since cleavage of the γ subunit into its most modified form (γ^3) occurs during the secretion of 7S NGF into saliva (BURTON et al. 1978), this processing may increase the stability of the complex in saliva.

The γ subunit has been implicated in the processing of the β-NGF precursor (see Sect. 3). The C-terminal amino acid sequence of the β-NGF precursor is Arg-Arg-Gly. The γ subunit cleaves the bond between the two Arg residues, liberating a dipeptide, and remains bound to the C-terminal Arg residue of β-NGF (BERGER and SHOOTER 1977; BOTHWELL and SHOOTER 1978). Removing the C-terminal Arg with carboxypeptidase B prevents the γ subunit from binding to β-NGF and thus prevents formation of the 7S complex (MOORE et al. 1974). It has been postulated, therefore, that the γ subunit binds to the C-terminal of β-NGF at its active site.

It is unclear whether in vivo the γ subunit cleaves only the C-terminal dipeptide of β-NGF or is involved in the N-terminal processing of β-NGF as well, since both the γ subunit and trypsin are able to correctly process the N-terminal end of the β-NGF precursor (see Sect. 4). It is also unclear whether other enzymes process the β-NGF precursor, since the γ subunit has not been found in most tissues.

2.1.3 The α Subunit

The complete nucleotide sequence of the α subunit demonstrates that there is significant homology between the α and γ subunits and reveals that the α subunit is a glandular kallikrein as well (EVANS and RICHARDS 1985; ISACKSON et al. 1984). However, the α subunit lacks enzymatic activity, probably due to amino acid changes near the active site and at the N-terminal (ISACKSON and BRADSHAW 1984). Its function in the 7S NGF complex is unknown, although it may protect β-NGF from proteolytic degradation and may regulate its biological activity (see Sect. 5).

Like the γ subunits, the α subunits are glycosylated and are heterogeneous in isoelectric point. The origin of this heterogeneity has been less well-studied for the α subunits than for the γ subunits, but it appears that the α subunits are found only in the two chain form. A model whereby differential processing of the α subunit at a single region (residues 135–137) could produce four different two chain forms has been presented by ISACKSON et al. (1987). The α subunits are slightly more acidic than the γ subunits, ranging in isoelectric point from 4.1 to 4.6, whereas the isoelectric points of the γ subunits range from 5.2 to 5.8 (VARON and SHOOTER 1970). Unlike the γ subunits, the four forms of the α subunits show no differential binding to β-NGF (SMITH et al. 1968). There are two α subunits per 7S complex, making the stoichiometry of 7S NGF $\alpha_2\beta\gamma_2$ (SERVER and SHOOTER 1977).

2.2 High Molecular Weight NGF

The 7S complex is most stable between pH 5.0 and 8.0 but can be reversibly dissociated into its subunits by extremes of pH (VARON et al. 1968; SMITH et al. 1968). In the mouse submaxillary gland, 7S NGF may exist in equilibrium with significant amounts of dissociated subunits (VARON and SHOOTER 1970; PANTAZIS and JENSEN 1985) and with various forms of less than 7S (YOUNG et al. 1978). In fact, the first NGF complex, purified by COHEN in 1960, had a sedimentation coefficient of 4.3S. Although its subunit composition has not been elucidated, SERVER and SHOOTER demonstrated that, in the absence of γ subunits, an $\alpha_2\beta$ complex can form (1976). YOUNG et al. (1988) have isolated a submaxillary gland high molecular weight NGF with a stoichiometry of $\alpha_2\beta\gamma$. However, the biological significance, if any, of these other forms of high molecular weight NGF has not been established.

Several other factors besides pH influence the stability of the high molecular weight NGF complex. For example, 7S NGF contains one or two zinc ions that stabilize the complex (PATTISON and DUNN 1975; BOTHWELL and SHOOTER 1977). The zinc may bind to the γ subunit in the complex, since it has been shown to inhibit the activity of the γ subunit. Also, YOUNG et al. (1988) demonstrated that a complex containing only one γ subunit contains only one zinc ion.

Whereas 7S NGF is isolated as a complex in which the α and γ subunits protect the β subunit from proteolysis, 2.5S NGF is isolated by dissociation of the 7S complex prior to the end of the purification procedure, which encourages proteolysis of the β subunit by contaminating proteases and carboxypeptidases (BOCCHINI and ANGELETTI 1969; ANGELETTI et al. 1973). 2.5S NGF consists of a mixture of molecules modified at both N- and C-terminals (see Sect. 4 for details of these modifications). Proteolysis at the C-terminal end of β-NGF prevents reassociation of the subunits into the 7S complex.

Binding of α and γ subunits to β-NGF regulates its biological activity in mouse submaxillary gland and saliva by influencing precursor processing, protecting β-NGF from proteolytic degradation, and inhibiting receptor binding (EDWARDS et al. 1988a; NICHOLS and SHOOTER 1985; HARRIS-WARRICK et al. 1980). The evidence for the existence of NGF binding molecules in other species and tissues (see below) and the potential importance of this means of regulation suggest that NGF binding molecules should be studied in other species. These regulatory molecules may or may not be α- or γ-like proteins.

3 Sources of NGF

There are few sources of NGF sufficiently rich for purification and molecular experimentation. The traditional source has been the submaxillary gland of the male mouse. Fighting among male mice results in the discharge of large amounts of salivary NGF into the bloodstream, suggesting that NGF may be important in the fighting mechanisms of male mice (ALOE et al. 1986). However, the salivary glands of other animals do not synthesize high levels of NGF, so the function of mouse submaxillary gland NGF in this tissue remains speculative. Reasonably high levels of NGF are contained in several other exocrine tissues and in their secretions, including snake venom (COHEN 1959; ANGELETTI 1968; HOGUE-ANGELETTI et al. 1976; BAILEY et al. 1975), guinea pig prostate (CHAPMAN et al. 1981; RUBIN and BRADSHAW 1981), bull semen and seminal vesicles (HARPER et al. 1982; HOFFMAN and UNSICKER 1982), and the submaxillary gland of the African rat, *Mastomys natalensis* (ALOE et al. 1981; DARLING and FAHNESTOCK 1988). The physiological function of NGF from these sources is also unknown.

NGF from other (endocrine) sources is synthesized in much smaller amounts than in the exocrine tissues and, therefore, with few exceptions, NGF from nonexocrine sources has been characterized only at the nucleic acid level.

Isolation of cDNA clone coding for mouse NGF (Scott et al. 1983; Ullrich et al. 1983) led to the cloning of bovine, avian, rat, guinea pig, *Mastomys*, and human NGF genes (Meier et al. 1986; Ebendal et al. 1986; Wion et al. 1986; Whittemore et al. 1988; Schwarz et al. 1989; Fahnestock and Bell 1988; Ullrich et al. 1983), demonstrating excellent conservation of the NGF sequence among divergent species.

This homology contrasts with the poor immunological cross-reactivity among NGFs of different species. In some cases, antibodies to mouse NGF have shown no cross-reactivity with NGF from chick, bull, and human (Bailey et al. 1976; Server and Shooter 1977; Walker et al. 1980; Beck and Perez-polo 1982; Harper and Thoenen 1981). In other cases, at least partial cross-reactivity was obtained between anti-mouse β-NGF and the NGF from *Mastomys*, human glial cells, guinea pig, and chick (Darling and Fahnestock 1988; Norrgren et al. 1980; Chapman et al. 1981; Belew and Ebendal 1986). Although the variable cross-reactivity is puzzling for such a highly conserved molecule, the presence of clusters of nonconserved residues located in potentially antigenic regions of an otherwise highly conserved molecule may be responsible (Meier et al. 1986; Whittemore et al. 1988). Alternatively, molecules that bind NGF in vivo may interfere with the binding of antibodies to the β subunit, thus preventing inhibition of biological activity when the bioassay is performed with partially purified NGF.

There is evidence that NGF from other tissues or sources may be found as high molecular weight complexes, but there is little, if any, evidence for α- or γ-like subunits in any tissues other than mouse and *Mastomys* submaxillary glands and snake venom gland.

3.1 Mastomys NGF

Mastomys natalensis is a South African rodent whose submaxillary glands contain high levels of NGF (Aloe et al. 1981). High molecular weight NGF can be purified from both male and female *Mastomys* (Darling and Fahnestock 1988). This material exhibits nerve growth-promoting activity comparable to that of the mouse, but it sediments at 5.1S, whereas the mouse high molecular weight complex sediments at 7S. Moreover, the 5S *Mastomys* complex exhibits no esterase activity, suggesting that this complex lacks γ subunits. The two types of subunit in the 5S complex migrate identically to mouse β and α subunits on isoelectric focusing gels and SDS–PAGE.

The sequence of the β-NGF cDNA of *Mastomys* is highly homologous to that of the mouse (Fahnestock and Bell 1988), and antibodies raised in rabbits against mouse β-NGF cross-react with *Mastomys* β-NGF (Darling and Fahnestock 1988). However, *Mastomys* β-NGF is resistant to carboxypeptidase B digestion (Fig. 1), which in the mouse removes the C-terminal arginine and prevents binding of the γ subunit. This resistance, as well as the lack of a γ subunit in the 5S complex, raised questions about the structure of the C-terminal of *Mastomys* β-NGF. The C-terminal sequence of the *Mastomys*

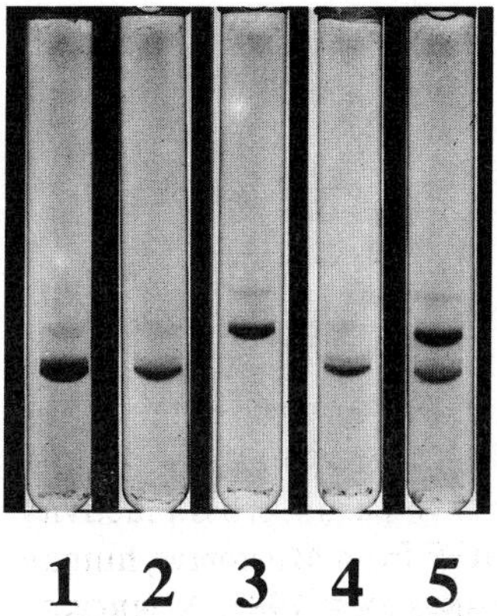

Fig. 1. Carboxypeptidase B digestion of mouse and *Mastomys* β-NGF. Quantities of 10 μg mouse and *Mastomys* β-NGF were dried, solubilized in buffer containing 2 *M* urea, and digested with carboxypeptidase B. Digests were analyzed by isoelectric focusing in the presence of urea. *Lane 1* and *Lane 2* are controls of mouse and *Mastomys* β-NGF incubated in the absence of enzyme. *Lane 3* and *Lane 4* show mouse and *Mastomys* β-NGF after treatment with carboxypeptidase B. *Lane 5* shows the result of digestion of both molecules in the same reaction mixture

β subunit precursor cDNA is Pro-Arg-Arg-Gly, similar to the mouse sequence, Thr-Arg-Arg-Gly (Fahnestock and Bell 1988). The *Mastomys* β-NGF precursor, like the mouse β-NGF, should be cleavable by a γ-like enzyme. *Mastomys* and mouse β-NGFs show identical charge properties on isoelectric focusing gels, supporting the conclusion that this dipeptide is cleaved in mature *Mastomys* β-NGF. However, this cleavage results in the presence of a penultimate proline residue, which we postulate is responsible for the resistance of *Mastomys* β-NGF to carboxypeptidase B and which may reduce the binding affinity of a γ-like enzyme to the C-terminal of *Mastomys* β-NGF.

The identity of the second subunit of the *Mastomys* 5S complex is less clear. The lack of esterase activity in the 5S complex indicates that the second subunit is α-like rather than γ-like. Although it comigrates with mouse α subunit on isoelectric focusing gels, antibodies raised against mouse α subunit do not cross-react with the *Mastomys* subunit nor do specific oligonucleotides directed against the N-terminal of mouse α subunit hybridize to *Mastomys* mRNA (Darling and Fahnestock 1988).

The *Mastomys* 5S complex is more difficult to dissociate than the mouse 7S complex. Mouse 7S NGF dissociates at pH 4.0, whereas *Mastomys* 5S NGF will not dissociate under these conditions but instead will dissociate at pH 3.0 in the presence of urea (Darling and Fahnestock 1988). The role of zinc ions has been studied in the stabilization of the mouse 7S complex (Bothwell and Shooter 1978; Palmer and Neet 1980) but not in the *Mastomys* high molecular weight complex. Since zinc ions are thought to bind primarily through the γ subunit, it will be of interest to determine whether zinc is important in stabilizing a complex that lacks the γ subunit.

3.2 Snake Venom NGF

Snake venom NGF has been isolated as a β-γ complex from *Crotalus adamenteus* (Perez-Polo et al. 1978). Unfortunately, the amino acid sequence of the

C-terminal portion of *Crotalus* β-NGF has not been determined. A high molecular weight complex has also been reported for NGF from *Vipera russelli*; in this case, no γ activity was detectable (PEREZ-POLO et al. 1983). A β-NGF cDNA from *Naja naja* (cobra) has been sequenced completely (SELBY et al. 1987a), but a high molecular weight NGF has never been isolated from this species. The C-terminal amino acid sequence suggests a reason: the amino acid sequence of *Naja naga* β-NGF, as determined from the nucleic acid sequence, is Lys-Lys-Gly-Asn, which might be cleavable by a kallikrein at the Lys residues. However, sequence analysis of the protein isolated from the venom glands showed that the C-terminal residue is asparagine (HOGUE-ANGELETTI et al. 1976) indicating that the C-terminal cleavage does not take place in vivo. It should be noted that neither mouse γ nor α subunits will bind to *Naja naja* NGF (SERVER et al. 1976). This may be due to the lack of cleavage at the Lys residues as well as to the requirement of mouse γ-NGF for a C-terminal arginine residue.

3.3 Human NGF

Due to the extremely low amounts of NGF synthesized in human tissues, the study of human NGF is very difficult. Many studies using competitive immunoglobulin or receptor binding assays have reported conflicting NGF levels in human serum. Furthermore, all are probably incorrect, inasmuch as this assay measures competition by human serum components for NGF binding sites rather than the binding competition between mouse and human NGF (THOENEN and BARDE 1980). Human NGF has been isolated from placenta and was shown to exhibit partial immunological cross-reactivity with anti-mouse β-NGF (GOLDSTEIN et al. 1978). The exact identity of the material is in doubt, since antibodies to mouse NGF have not been shown to inhibit the biological activity of the human material (WALKER et al. 1980; BECK and PEREZ-POLO 1982). However, NGF activity has been detected in human glial cells in culture, and this activity is blocked by anti-mouse β-NGF (NORRGREN et al. 1980).

Despite the difficulty of studying the human NGF protein, it is clear that NGF is synthesized in human tissues: the presence of NGF mRNA has been demonstrated in human placenta, cortex, and hippocampus, with very low levels in striatum and septum/nucleus basalis (COLE et al. 1986; GOEDERT et al. 1986). The human NGF gene has been clone, and nucleotide sequence analysis demonstrates that mouse and human NGF genes are highly homologous (ULLRICH et al. 1983). Human NGF has a C-terminal sequence, Arg-Arg-Ala, that is similar to the mouse sequence, Arg-Arg-Gly, and thus could be cleaved by a γ subunit. However, although there is conflicting evidence regarding γ-like activity in human placental NGF, no γ subunit gene or protein has so far been isolated from human tissues. Nonetheless, it appears that human placental NGF may be complexed with an α subunit (BLUM et al. 1981; PEREZ-POLO et al. 1983).

3.4 Other Sources of NGF

In the early stages of purification, guinea pig prostate NGF behaves as a high molecular weight complex. Chapman et al. (1981) reported that this complex exhibits esterase activity, whereas Rubin and Bradshaw did not detect any esterase activity in their high molecular weight complex (1981). The isoelectric focusing pattern of guinea pig β-NGF suggests a partial cleavage at the C-terminal, analogous to that of mouse β-NGF (Chapman et al. 1981). Furthermore, the C-terminal sequence of guinea pig NGF is similar to mouse NGF in that it ends in Arg-Arg-Gly (Schwarz et al. 1989). The primary source of esterase activity in guinea pig prostate does not appear to form a complex with β-NGF (Dunbar and Bradshaw 1985). This does not rule out the existence of α- or γ-like subunits in the guinea pig complex, however.

In contrast, the C-terminals of both bovine and chick NGF differ from the mouse sequence in a way which suggests that these NGFs cannot be processed by a γ subunit to form a β–γ complex (Ebendal et al. 1986; Meier et al. 1986). The bovine NGF C-terminal sequence is Thr-Gly-Gln and the chicken NGF C-terminal sequence is Ser-Gly-Pro, neither of which would bind to the active site of a γ-like enzyme. However, chick embryo NGF is thought to be associated with other, unidentified proteins (Belew and Ebendal 1986), and bovine NGF purified from seminal vesicles and seminal plasma is found in a high molecular weight form (Hoffman and Unsicker 1982; Harper et al. 1982). The components of the chick and bovine high molecular weight complexes have not been studied.

Many cells in culture, including fibroblasts and muscle and glial cells, synthesize NGF (Bradshaw and Young 1976; Pantazis 1983; Siminoski and Murphy 1987; Murphy et al. 1977a, b; Norrgren et al. 1980; Furukawa et al. 1984). In addition, several cell types, such as the original S-180 sarcoma cells used by Levi-Montalcini and Hamburger (1953), rat muscle cells, and mouse fibroblasts, synthesize high molecular weight NGF (Kim and Pantazis 1985; Siminoski and Murphy 1987; Pantazis et al. 1977; Murphy et al. 1977b). The high molecular weight form of NGF from mouse S-180 cells has been shown to contain α and γ subunits (Werrbach-Perez and Perez-Polo 1987), but mouse fibroblasts appear to contain β-NGF-binding molecules other than α-NGF and γ-NGF (Pantazis 1983; Siminoski and Murphy 1987).

3.5 Comparison of High Molecular Weight NGFs

Interestingly, all mammalian high molecular weight NGF complexes are more difficult to dissociate than is mouse 7S NGF. Mouse 7S NGF is dissociated at pH 4; guinea pig and bovine high molecular weight NGF dissociate at pH 3 (Chapman et al. 1981; Harper et al. 1982); human placental NGF requires low pH and 5 *M* urea for dissociation (Blum et al. 1981; Perez-Polo et al. 1983); and

Mastomys NGF dissociates at pH 3.0 in the presence of urea (DARLING and FAHNESTOCK 1988). High molecular weight NGFs purified from mouse fibroblasts and rat muscle cells in culture are much more stable under dilute conditions than is submaxillary gland 7S NGF (PANTAZIS et al. 1977; MURPHY et al. 1977b), indicating that the subunit contacts may differ significantly among mouse 7S NGF and other high molecular weight NGFs. Interestingly, a mouse submaxillary gland $\alpha_2\beta\gamma$ complex purified by YOUNG et al. (1988) is also stable under dilute conditions. It should be noted that failure to dissociate a high molecular weight complex may lead to the erroneous conclusion that it contains no γ-esterase activity, since this activity is inhibited in the complex.

Although the β-NGFs from a number of sources have been compared at both the protein and nucleic acid levels, it is clear that much work remains to be done to clarify the differences between the high molecular weight NGFs. The complexes from a variety of mammalian tissues and cell types differ in their properties. The identification of the non-β components of these complexes will require a great deal more work.

4 NGF Biosynthesis

4.1 Transcriptional Regulation

β-NGF is synthesized in the granular convoluted tubule cells of the submaxillary gland of the adult male mouse (CARAMIA et al. 1962; SCHWAB et al. 1976). The α and γ subunits have been shown to be present in the same cells (MOWRY et al. 1984). All three subunits of submaxillary gland 7S NGF are androgen-induced (ISACKSON et al. 1987) so that the male submaxillary gland produces larger amounts of NGF than does the female gland. This difference has been reported to be on the order of 10-fold (CARAMIA et al. 1962; SCOTT et al. 1983) to 30-fold (ISACKSON et al. 1987), and it occurs at the level of transcription. Furthermore, NGF levels are decreased in castrated male mice and are increased in testosterone-treated female mice (CARAMIA et al. 1962; ISHII and SHOOTER 1975). NGF in mouse submaxillary gland is also inducible by thyroxine (ALOE and LEVI-MONTALCINI 1980).

The difference in the amount of NGF synthesized between male and female in mouse submaxillary gland does not, in general, occur in other species or tissues. Relatively high levels of NGF are found in male and female reproductive organs of several species (PEREZ-POLO et al. 1983; HARPER et al. 1982) and NGF synthesis is inducible by testosterone in mouse vas deferens (GOLDSTEIN et al. 1989). There is meager or conflicting evidence for androgen inducibility of β-NGF in other nonreproductive mouse tissues, including adrenals, superior cervical ganglia, and brain (KORSCHING and THOENEN 1988; KATOH-SEMBA et al. 1989; SELBY et al. 1987b), and no evidence for androgen inducibility in nonreproductive tissues of other species, including rat submaxillary gland (HEUMANN et al. 1984). No difference in β-NGF levels is found in the *Mastomys* submaxillary gland; male and female glands contain approximately the same high concentrations of β-NGF (ALOE et al. 1981; BOWCOCK et al. 1988).

NGF is also regulated developmentally (LARGE et al. 1986) and in response to injury and tissue explantation (MILLARUELO et al. 1986; SHELTON and REICHARDT 1986a; HEUMANN et al. 1987). Levels of β-NGF protein are correlated with the amounts of β-NGF mRNA in a variety of tissues (HEUMANN et al. 1984; SHELTON and REICHARDT 1986b). Furthermore the expression of a NGF growth hormone fusion gene in transgenic mice demonstrates the expected tissue specificity (ALEXANDER et al. 1989). Thus, the concentration of NGF in target tissues is generally regulated at the level of transcription.

4.2 Transcription of β-NGF

The mouse β-NGF gene covers more than 43 kb and consists of five exons separated by four introns (Selby et al. 1987b). The NGF gene is present as a single copy in both the mouse and human genomes and has been localized to chromosome 3 in mouse and to the short arm of chromosome 1 in human (ZABEL et al. 1985; FRANCKE et al. 1983). The rat and mouse NGF promoter regions have been identified (ZHENG and HEINRICH 1988); our laboratory has obtained preliminary sequence data for the human NGF promoter region.

It was shown recently that there are two major and two minor β-NGF transcripts in mouse (SELBY et al. 1987b). The two major transcripts differ as a result of alternative RNA splicing (EDWARDS et al. 1986). One major transcript (transcript A, coding for the "long" precursor) is found primarily in the mouse submaxillary gland and in the placenta from several species, whereas the other major transcript (transcript B, coding for the "short" precursor) is the predominant form in other tissues. The two precursors contain the same NGF coding sequence near their 3′ ends. They differ, however, at their 5′ ends in that the longer transcript encodes a 34 000 molecular weight (MW) precursor containing a hydrophobic signal peptide 70 residues downstream from the initiation methionine, whereas the shorter transcript encodes a 27 000 MW precursor containing its hydrophobic signal peptide at the N-terminal end.

The two minor transcripts are similar to the major ones. Transcript C is similar to transcript A but is derived from an independent promoter slightly upstream of the transcript A promoter; transcript D is similar to transcript B but appears to have an intron spliced out and may represent a partially processed transcript. In mouse submaxillary gland, transcript A is present in threefold excess over transcript B, whereas in most other tissues transcript B is present in fourfold excess over transcript A. In submaxillary gland, transcripts C and D comprise less than 1% of NGF mRNA, but in cortex and heart the level of transcript C is 5%–10% of the total NGF mRNA. In other tissues, both transcripts are barely detectable. The levels of the major transcripts change in parallel during postnatal development, and all transcripts contain the same mature NGF sequence at their 3′ ends; therefore the function of several precursors and the reasons for the predominance of particular forms in various tissues are not understood.

4.3 Translation of β-NGF

The β-NGF protein is synthesized as a large precursor (prepro-NGF) that is processed at both N- and C-terminal ends into the mature 13 200 MW protein. The long mRNA from mouse submaxillary gland (transcript A) uses a methionine residue at −187 as the initiator methionine, producing a precursor with a MW of 33 800; the short mRNA (transcript B), which is spliced in such a way as to delete methionine −187, uses a methionine at residue −121, producing a precursor with a MW of 27 000. Three potential glycosylation sites are present in the β-NGF precursor: two are present in the N-terminal region of pre-β-NGF but not in the mature protein, and one is present in the mature 118 amino acid β-NGF sequence. Although the glycosylation site is only partially utilized in mature β-NGF (MURPHY et al. 1989), it is not known whether any or all of the sites in the precursor are glycosylated during biosynthesis.

4.4 Precursor Processing

It has proven difficult over the years to study processing of the β-NGF precursor using immunoprecipitation of processed intermediates because antibodies directed against mature NGF are seldom able to precipitate the NGF precursor or its intermediates, especially those synthesized by cell-free translation systems (THOENEN and BARDE 1980; DICOU et al. 1986). Furthermore, although the cDNA sequence of mouse β-NGF predicts three new peptides by virtue of processing of the precursor at pairs of basic amino acids, generally not all such sites are used and other, less obvious sites are sometimes cleaved.

However, a relatively stable NGF precursor of molecular weight 22 000 was reported by BERGER and SHOOTER (1977). Using several different antibody preparations, DARLING et al. (1983) identified a series of β-NGF precursors including 35 000, 29 000, 22 000, 19 000, and 13 000 MW species. The 35 000 MW species almost certainly represents prepro-NGF as synthesized from the large transcript A; the 29 000 MW species, although identified by DARLING et al. as perhaps being pro-NGF, could be prepro-NGF synthesized from the small transcript B. The 22 000 and 19 000 MW species probably represent processing intermediates arising from cleavage of the NGF precursor at pairs of basic amino acids, specifically Arg-Arg at −65 and −66 and Lys-Lys-Arg-Arg at residues −41 to −38.

The mature β-NGF arises from cleavage following a Lys-Arg (−2 to −1) at the N-terminal end and between Arg-Arg at residues 118-119 at the C-terminal end. The order in which processing takes place and the biological activity, if any, of the cleaved peptides have not been determined.

4.5 The Role of the γ Subunit

The second subunit of the 7S NGF complex, the γ subunit, has been implicated in the processing of the β-NGF precursor (BERGER and SHOOTER 1977). The γ subunit is thought to cleave a C-terminal dipeptide from β-NGF, which exposes an arginine residue and thus permits the γ subunit to bind to β-NGF in the complex. Removal of the C-terminal arginine of mature β-NGF prevents binding of the γ subunit in the 7S complex (GREENE et al. 1968; MOORE et al. 1974; BOTHWELL and SHOOTER 1978).

It also has been proposed that the γ subunit performs the cleavages required for processing of the N-terminal of pro-NGF. BERGER and SHOOTER (1977) demonstrated cleavage of the 22000 MW precursor species to NGF by the γ subunit. Although the EGF binding protein also was capable of carrying out these same cleavages, the authors interpreted this result cautiously because the precursor probably does not retain its native conformation when isolated as an immunoprecipitate or after reduction with β-mercaptoethanol in urea and denaturation in SDS. In addition, EGF binding protein will not form a complex with native β-NGF (SERVER and SHOOTER 1976).

The γ subunit also has been shown to generate 22000 and 18000 MW intermediates as well as a small amount of mature NGF from 34000 and 29000 MW precursors produced in vitro by wheat germ extracts (JONGSTRA-BILEN et al. 1989). The low yield was attributed to the improper conformation of the NGF precursor.

EDWARDS et al. (1988a, b) have attempted to solve the conformation question by expressing the native β-NGF precursor in mammalian cells using a vaccinia virus system. They were able to show that both the long and short precursors are cleaved to mature β-NGF by stoichiometric amounts of the γ subunit. They also showed that trypsin in catalytic quantities produces properly processed mature NGF and that the presence of α-NGF and zinc did not influence β-NGF precursor cleavage by the γ subunit, raising the question of whether this system adequately tests for the postulated in vivo specificity. However, the ability of a number of different cell lines, which do not necessarily contain γ subunits, to accurately process the β-NGF precursor calls into question the specificity of processing by γ-NGF. It may be that in vivo the specificity of processing is determined more by enzyme localization than by substrate specificity.

4.6 Biosynthesis of the γ Subunit

The biosynthesis of the γ subunit has not been studied in detail. It is synthesized from the *mGK-3* gene (EVANS and RICHARDS 1985) as a single chain zymogen requiring cleavage of an N-terminal peptide for activation, much like chymotrypsin. As discussed in Sect. 2, the γ subunit is activated and cleaved internally to create two and three chain forms. The nature of the enzymes that perform the zymogen activation and the chain cleavages is unknown, although the amino acid sequences of the cleavage sites suggest trypsin-like enzymes, perhaps other kallikreins.

4.7 Biosynthesis of the α Subunit

The α subunit is synthesized from the *mGK-4* gene (Evans and Richards 1985). The prepro-α subunit migrates as a single chain at 30 000 MW, whereas the pro-α subunit, which is glycosylated, migrates at a MW of 32 000 (Darling et al. 1983). The pro-α subunit is cleaved at two sites to produce the mature α subunit, which consists of 16 000 and 10 000 MW chains linked by disulfide bridges, and exists in multiple forms that differ in charge. A model (Sect. 2) has been suggested to explain the existence of these forms (Isackson et al. 1987). As with the γ subunit, the enzymes that perform the chain cleavages are unknown.

Unlike the γ subunit, the α subunit does not undergo zymogen activation, most likely due to an Arg-to-Gln substitution at residue -1 and perhaps also due to a deletion of four N-terminal residues (Isackson and Bradshaw 1984; Isackson et al. 1984). The inability to undergo this cleavage and subsequent conformational change may account for the α subunit's lack of esteropeptidase activity. However, there is also a substitution of a glycine residue for a histidine near the active site serine and an Asp to Tyr substitution in the binding pocket. Whether the active site substitutions, the N-terminal changes, or both contribute to the α subunit's lack of activity is unclear.

4.8 The Role of Other Kallikreins in NGF Processing

The α and γ subunits belong to a large family of trypsin-like serine proteases known as glandular kallikreins. In the mouse, the kallikreins comprise a family of 14 active genes and 10 pseudogenes (Evans et al. 1987), all highly homologous, with at least 10 of the 14 active genes expressed in the submaxillary gland. Each of the kallikreins displays amino acid differences in the active site region, suggesting that each kallikrein has a specific substrate or substrates. The argument for substrate specificity is strengthened by experiments demonstrating that EGF binding protein and γ-NGF—two very similar kallikreins—are unable to substitute for each other in their respective complexes (Server and Shooter 1976). Further evidence for strict substrate specificity of these molecules is the inability to assemble complexes of mouse and *Crotalus* subunits (Perez-Polo et al. 1978). Therefore, the experiments of Edwards et al. (1988a, b) notwithstanding, it is unclear whether in vivo the γ subunit carries out all the cleavages required for the maturation of the β-NGF precursor or whether other trypsin-like enzymes participate in this processing in submaxillary gland or other tissues.

The study of the biosynthesis and substrate specificities of the various kallikreins, particularly the α and γ subunits, is complicated by the existence of the very large and highly homologous kallikrein family in the mouse and by the synthesis of many of these enzymes in the mouse submaxillary gland. The mRNAs and the protein precursors of these enzymes migrate similarly or identically on gels, and antibodies to one frequently immunoprecipitate other members of the family as well. In tissues other than the mouse submaxillary gland, these proteins often are present in extremely small amounts, if at all. It is difficult therefore to determine whether, in tissues that contain β-NGF, the α and γ subunits also are present.

Furthermore, it is not clear that the γ subunit is involved in the C-terminal processing of the β-NGF precursor in tissues other than the submaxillary gland. Although the presence of β-NGF has been demonstrated, γ subunits have not been found in either denervated rat iris or mouse fibroblast cells in culture (MURPHY et al. 1986; PANTAZIS 1983). Other investigators, who claim to have identified α and/or γ subunits in tissues containing β-NGF (PANTAZIS and JENSEN 1985), particularly in studies using immunoprecipitation, often have not addressed the possibility that they have identified another closely related kallikrein. Therefore, the apparent lack of a γ subunit in many β-NGF-synthesizing tissues and the ability of other enzymes and a variety of cell lines to correctly process the β-NGF precursor suggest that other enzymes may function in NGF processing in different species and tissues.

4.9 Biosynthesis of NGF from *Mastomys natalensis*

The functions of the various subunits in the mouse high molecular weight complex and the pathways of biosynthesis of these subunits have been difficult to illuminate. This is especially true for the α subunit, which has no detectable enzymatic or biological activity. The presence of a β-γ complex in *Crotalus* snake venom affords the opportunity to study a complex in the absence of the α subunit; *Mastomys* allows the study of a complex lacking tightly bound γ subunits. The differences between these high molecular weight NGF complexes therefore allow us a unique opportunity to study the functions and the importance of all three types of subunits involved in NGF action.

A cDNA for *Mastomys* β-NGF has been isolated and sequenced (FAHNESTOCK and BELL 1988). There are four possible initiation codons for the *Mastomys* β-NGF open reading frame. No information exists on the number of β-NGF precursors in *Mastomys*, but hydropathicity analysis (KYTE and DOOLITTLE 1982) suggests that both methionines -183 and -121 may serve as initiation codons. By analogy with the mouse (EDWARDS et al. 1986), the *Mastomys* gene would encode at least two precursors, a short and a long. The 5′

region of *Mastomys* β-NGF cDNA (the sequence upstream of base 157) diverges significantly from the mouse sequence; thus, although the short β-NGF precursor of *Mastomys* should be very similar to the mouse short precursor, the long precursors should differ significantly at their N-terminals.

The coding region and the 3′ untranslated region of *Mastomys* β-NGF cDNA share 95% homology with mouse β-NGF cDNA. Most major structural features, such as histidine and tyrosine residues essential for biological activity, cysteine residues, N-linked glycosylation sites, and potential proteolytic cleavage sites, are conserved between mouse and *Mastomys*. In fact, only 40% of the nucleotide differences between the mature β-NGFs of *Mastomys* and mouse result in amino acid changes. As in the mouse cDNA, the terminal sequence of the 3′ untranslated region of *Mastomys* β-NGF is exceptionally A + T rich and contains a polyadenylation signal.

The *Mastomys* β-NGF precursor has the same three C-terminal residues as the mouse, suggesting that the *Mastomys* β-NGF precursor could interact with a γ-like subunit (see Sect. 2). By analogy with the mouse, a cleavage following a conserved Lys-Arg at the N-terminal and at the conserved C-terminal site by a γ-like enzyme would result in a 118 residue, mature *Mastomys* β-NGF. This choice of cleavage sites is supported by data demonstrating comigration of *Mastomys* β-NGF with mouse β-NGF on SDS-PAGE (Darling and Fahnestock 1988).

The γ subunit in mouse has a function beyond that of dipeptide cleavage; when complexed to β-NGF, it protects β-NGF from further degradation (Moore et al. 1974). The lack of a tightly bound γ subunit in *Mastomys* may have resulted in compensating changes that make the *Mastomys* β-NGF less susceptible to degradation. Little is known about the processing of NGF precursors in mouse or *Mastomys*. The kallikrein family in *Mastomys* is very similar in size and sequence homology to the mouse kallikrein family (Bowcock et al. 1988). There are many kallikreins expressed in the *Mastomys* submaxillary gland that could perform the necessary processing steps (Bowcock et al. 1988; Darling and Fahnestock 1988). Therefore, the γ subunit or γ-like proteins could perform the processing steps required for the maturation of *Mastomys* β-NGF, but may not remain bound as part of the high molecular weight complex. Alternatively, the γ subunit could be bound more loosely in the *Mastomys* complex than is the mouse γ subunit in the 7S complex and thus could be lost more easily during the purification procedure. A report of a *Mastomys* high molecular weight complex that exhibits esterase activity (Burcham and Shooter 1987) supports the latter hypothesis. A study of the processing steps involved in β-NGF maturation in a variety of tissues and the nature of the subunits in other high molecular weight complexes will shed some light on these same questions in the mouse 7S complex.

5 Proteolysis of Mature β-NGF

Murine β-NGF can be stored in the submaxillary gland or secreted into the saliva. It is present as 7S NGF in both gland and saliva (BURTON et al. 1978), where the environment contains many proteolytic enzymes that can degrade NGF. The α and γ subunits of the 7S complex partially protect β-NGF from degradation (MOORE et al. 1974; MOBLEY et al. 1976; NICHOLS and SHOOTER 1985). However, a number of proteolytic modifications commonly are found in NGF purified from either submaxillary gland tissue or saliva, supporting the hypothesis that 7S NGF exists in equilibrium with biologically significant amounts of dissociated forms. Degradation of the mature β subunit can occur at both N- and C-terminal ends.

5.1 Processing at the N-Terminal

At the N-terminal of mouse β-NGF, the histidine-methionine site at residues 8-9 is cleaved by the enzyme β-NGF-endopeptidase to liberate an N-terminal octapeptide from mature β-NGF (ANGELETTI et al. 1973). Both the α and γ subunits, when bound to β-NGF in the 7S complex, prevent this cleavage (MOBLEY et al. 1976; NICHOLS and SHOOTER 1985). β-NGF-endopeptidase is a kallikrein and in the mouse it is coded for by the gene *mGK-22* (FAHNESTOCK et al. 1988; FAHNESTOCK et al., manuscript submitted). The cleavage carried out by this enzyme does not appear to influence NGF regulation; des(octa)β-NGF exhibits biological activity and it will form the 7S complex as readily as full-length NGF (MOBLEY et al. 1976). Although the octapeptide does not exhibit any NGF activity (HOGUE-ANGELETTI et al. 1974), recent studies indicate that it exhibits hyperalgesic activity (LEVINE et al. 1988; TAIWO et al., manuscript submitted).

In human, bovine, and chicken NGF, the site at residue 8-9 is highly conserved as either His-Met or His-Arg (HARPER et al. 1982). The His-Arg site also should be susceptible to a kallikrein-like enzyme, since these enzymes generally cleave at basic residues such as lysine or arginine. However, no des(octa)NGF has been reported for these three NGFs—possibly because the des(octa)NGF was not sought or because the NGF was purified from tissues other than submaxillary gland. The tissue distribution of β-NGF-endopeptidase has not been studied. In mouse, β-NGF's existence in the des(octa) form in tissues other than submaxillary gland is not known.

Mastomys β-NGF contains a glutamine residue at position 8 in contrast to the histidine of mouse. The sequence change suggests that the β-NGF-endopeptidase cleavage cannot take place, even in the absence of bound γ subunit. This hypothesis is supported by the lack of heterogeneity exhibited by the *Mastomys* β-NGF: Whereas the mouse β-NGF consists of a mixture of full-length and des(octa) species, *Mastomys* NGF exhibits no heterogeneity when subjected to N-terminal amino acid sequencing. Moreover, the two forms present in mouse β-NGF are detectable on SDS–PAGE, but *Mastomys* β-NGF appears as a single species (DARLING and FAHNESTOCK 1988).

5.2 Processing at the C-Terminal

Another very common modification is the removal of the C-terminal arginine from the mouse β-NGF by a carboxypeptidase B-like enzyme. The removal of this residue prevents β-NGF from binding to the γ-subunit, because it is this arginine that is bound by the active site of the γ subunit (MOORE et al. 1974). However, this modification has no direct effect on the biological activity of NGF. It is the presence of the γ subunit, not the α subunit, that protects β-NGF from carboxypeptidase digestion (MOBLEY et al. 1976; NICHOLS and SHOOTER 1985). Once again, it has been shown from the nucleic acid sequence that other NGFs, such as bovine and human, have arginine residues capable of removal by a carboxypeptidase; the removal of this residue has only been demonstrated in mouse submaxillary gland, however.

Mastomys β-NGF has been shown to be carboxypeptidase B-resistant (Fig. 1). Originally this was thought to indicate the lack of a C-terminal arginine in this species, which could explain the lack of a tightly bound γ subunit in the *Mastomys* 5S complex. However, sequence analysis of the *Mastomys* β-NGF cDNA (FAHNESTOCK and BELL 1988) demonstrated the same C-terminal sequence as mouse but with a proline residue substituted for the threonine-117 in mouse. Both mouse and *Mastomys* β-NGF proteins show identical charge characteristics on isoelectric focusing gels, indicating that the most likely sequence for the C-terminal of mature *Mastomys* β-NGF is Pro-Arg-COOH (i.e., a processed form). This sequence should be resistant to carboxypeptidase B, whereas the mouse sequence, Thr-Arg-COOH, is susceptible to cleavage by this enzyme.

5.3 NGF in Blood and Saliva

Since no differences in biological activity between modified and unmodified NGF have yet been demonstrated (MOBLEY et al. 1976), the physiological significance of these proteolytic modifications has been unclear. The discovery of hyperalgesic activity by the octapeptide (LEVINE et al. 1988) provides a physiological reason for the N-terminal cleavage. NGF purified as 7S NGF from submaxillary gland tissue typically is modified at both N- and C-terminals in approximately 10% of the chains. In saliva, approximately 30% of the chains are present without an N-terminal octapeptide, while the C-terminal arginine has been removed in only 10% of the chains. The 2.5S NGF lacks the N-terminal octapeptide 50% of the time (SERVER and SHOOTER 1977).

In plasma, the NGF dimer is bound to α_2-macroglobulin (RONNE et al. 1979; KOO and STACH 1989), but the stage at which the NGF is dissociated from 7S and bound to α_2-macroglobulin and the occurrence of further proteolysis in plasma, if any, are unknown.

5.4 Regulation of NGF Activity in the 7S Complex

In addition to providing protection of β-NGF from proteolytic modification, the 7S complex also regulates the activity of the component subunits. In the 7S complex, the biological activity of NGF is suppressed (STACH and SHOOTER 1980). The mechanism of this suppression is through interference, by the α and γ subunits, with NGF binding to its receptor (HARRIS-WARRICK et al. 1980; NICHOLS and SHOOTER 1985). The esterase activity of the γ subunit is also inactivated by formation of the 7S complex (GREENE et al. 1968; BOTHWELL and SHOOTER 1978) due to the occupation of the γ subunit active site by the C-terminal arginine of β-NGF. Thus, β-NGF acts as an active site inhibitor of the γ subunit.

6 Future Prospects

6.1 The High Molecular Weight Complex

Mouse submaxillary gland NGF is problematic for a number of reasons. This is the only tissue other than *Mastomys* submaxillary gland and snake venom gland, both related tissues, in which NGF is found in such high quantities. It is also the only mammalian tissue in which the high molecular weight complex has been well characterized. There are two essential questions: Do the α and γ subunits function only in a storage capacity in this tissue? In other tissues with much lower levels of NGF, do these subunits exist or have any physiological significance?

It is possible that in tissues synthesizing small amounts of β-NGF, there are also small amounts of α- or γ-like subunits that may bind to NGF and function in its regulation. Clearly, the binding of such proteins to β-NGF can regulate its biological activity through the inhibition of receptor binding and can prevent the degradation of β-NGF by tissue proteases. The suggestion that other tissues, for example, bovine seminal vesicles, human placenta, chick embryo, and guinea pig prostate, may contain molecules that bind β-NGF leaves the role of high molecular weight NGF in regulating NGF activity uncertain for the time being.

6.2 Processing of β-NGF

Another question that currently remains unanswered is the role of the γ subunit in the processing of the β-NGF precursor. Although recent experiments demonstrate the ability of the γ subunit to correctly process pro-β-NGF at both N- and

C-terminal ends, the apparent lack of a γ subunit in many NGF-synthesizing tissues and the ability of trypsin to correctly process the β-NGF precursor suggest that other enzymes may function in NGF processing in different tissues.

Finally, the large family of kallikreins in the mouse, to which the α and γ subunits belong, is puzzling. The highly selective substrate specificities of these kallikreins, exemplified by the inability of the γ subunit and the EGF binding protein to substitute for each other in their respective complexes, led to the suggestion that these enzymes function in the highly selective processing of their hormone substrates. This finding contrasts with the fact that other species, humans in particular, appear to have a much smaller kallikrein family. Human kallikreins therefore must have broader substrate specificity, less redundancey than is proposed for the mouse, or must not function pivotally in the processing of a variety of hormone precursors. Furthermore, there may be a great deal of redundancy in the activity of the mouse kallikreins, with substrate specificity determined largely by tissue distribution. The answers to these and other questions may best be discovered by studying the biosynthesis, processing, and regulation of NGF in tissues and species other than the mouse submaxillary gland.

7 Summary

Most of our knowledge about NGF comes from extensive study of the mouse submaxillary gland protein. NGF from this source is isolated as a high molecular weight complex consisting of β-NGF and two subunits, α and γ, belonging to the kallikrein family of serine proteases. There are few other tissues where NGF is found in sufficient quantities for protein purification and study, although new molecular biological techniques have accelerated the study of NGFs from a variety of species and tissues.

Mouse submaxillary gland NGF is synthesized as a large precursor that is cleaved at both N- and C-terminals to produce mature NGF. This biologically active molecule can be further cleaved by submaxillary gland proteases. The roles of the α and γ subunits in the processing of the β-NGF precursor, the modulation of the biological activity of β-NGF, and the protection of mature β-NGF from degradation have been well studied in the mouse. However, the apparent lack of α and γ subunits in most other tissues and species and the existence of a large family of murine kallikreins, many of which are expressed in the submaxillary gland, challenge the relevance of murine high molecular weight NGF as a proper model for NGF biosynthesis and regulation. It is important therefore to identify and characterize other NGF complexes and to study their subunit interactions, biosynthesis, processing, and regulation.

This review points out a number of other species and tissues in which the study of NGF has just begun. At this time, there exist many more questions than answers regarding the presence and the functions of NGF processing and regulatory proteins. By studying NGF in other species and tissues and comparing the processing and regulation of NGF from several sources, we will discover the unifying concepts governing the expression of NGF biological activity.

Acknowledgements. I wish to thank Dr. Tom Darling for performing the carboxypeptidase B digestion of mouse and *Mastomys* NGF, and Rose Ann Bell for *Mastomys* NGF purification. Thanks also to Dr. Paul Johnson and Dot Timock for helpful reading of the manuscript, and to Gabriella Huff for manuscript preparation. This work was supported by NIH grant NS-23075.

References

Alexander JM, Hsu D, Penchuk L, Heinrich G (1989) Cell-specific and developmental regulation of a nerve growth factor-human growth hormone fusion gene in transgenic mice. Neuron 3: 133–137

Aloe L, Levi-Montalcini R (1980) Comparative studies on testosterone and L-thyroxine. Effects on the synthesis of nerve growth factor in mouse submaxillary salivary glands. Exp Cell Res 125: 15–22

Aloe L, Cozzari C, Levi-Montalcini R (1981) The submaxillary salivary glands of the African rodent *Praomys* (*mastomys natalensis*) as the richest available source of the nerve growth factor. Exp Cell Res 133: 475–480

Aloe L, Alleva E, Böhm A, Levi-Montalcini R (1986) Aggressive behavior induces release of nerve growth factor from mouse salivary gland into the bloodstream. Proc Natl Acad Sci USA 83: 6184–6187

Angeletti RH (1968) Studies on the nerve growth factor (NGF) from snake venom. Molecular heterogenity. J Chromatogr 37: 62–69

Angeletti RH, Hermodson MA, Bradshaw RA (1973) Amino acid sequences of mouse 2.5S nerve growth factor. II. Isolation and characterization of the thermolytic and peptic peptides and the complete covalent structure. Biochemistry 12: 100–115.

Bailey GS, Banks BEL, Pearce FL, Shipolini RA (1975) A comparative study of nerve growth factors from snake venoms. Comp Biochem Physiol 51B: 429–438

Bailey GS, Banks BEL, Carstairs JR, Edwards DC, Pearce FL, Vernon CA (1976) Immunological properties of nerve growth factors. Biochim Biophys Acta 437: 259–263

Beck CE, Perez-Polo JR (1982) Human β-nerve growth factor does not crossreact with antibodies to mouse β-nerve growth factor in a two-site radioimmunoassay. J Neurosci Res 8: 137–152

Belew M, Ebendal T (1986) Chick embryo nerve growth factor. Fractionation and biological activity. Exp Cell Res 167: 550–558

Berger EA, Shooter EM (1977) Evidence for pro-β-nerve growth factor, a biosynthetic precursor to β nerve growth factor. Proc Natl Acad Sci USA 74: 3647–3651

Blum M, Beck CE, Perez-Polo JR (1981) Subunit composition of NGF from human placenta at term. Soc Neurosci Abstr 7: 145

Bocchini V, Angeletti PU (1969) The nerve growth factor: purification as a 30,000-molecular-weight protein. Proc Natl Acad Sci USA 64: 787–794

Bothwell MA, Shooter EM (1977) Dissociation equilibrium constant of β-nerve growth factor. J Biol Chem 252: 8532–8536

Bothwell MA, Shooter EM (1978) Thermodynamics of interaction of the subunits of 7S nerve growth factor. The mechanism of activation of the esteropeptidase activity by chelators. J Biol Chem 253: 8458–8464

Bowcock AM, Fahnestock M, Goslin K, Shooter EM (1988) The NGF and kallikrein genes of mouse, the African rat *Mastomys natalensis* and man: their distribution and mode of expression in the salivary gland. Mol Brain Res 3: 165–172

Bradshaw RA, Young M (1976) Nerve growth factor: recent developments and perspectives. Biochem Pharmacol 25: 1445–1449

Burcham TS, Shooter EM (1987) The NGF complex from the African rat *Mastomys natalensis*. Soc Neurosci Abstr 13: 550

Burton LE, Wilson WH, Shooter EM (1979) Nerve growth factor in mouse saliva. J Biol Chem 253: 7807–7812

Caramia F, Angeletti PU, Levi-Montalcini R (1962) Experimental analysis of the mouse submaxillary salivary gland in relationship to its nerve-growth factor content. Endocrinology 70: 915–922

Chapman CA, Banks BEC, Vernon CA, Walker JM (1981) The isolation and characterization of nerve growth factor from the prostate gland of the guinea-pig. Eur J Biochem 115: 347–351

Cohen S (1959) Purification and metabolic effects of a nerve growth-promoting protein from snake venom. J Biol Chem 234: 1129–1137

Cohen S (1960) Purification of a nerve-growth promoting protein from the mouse salivary gland and its neuro-cytotoxic antiserum. Proc Natl Acad Sci USA 46: 302–311

Cohen S, Levi-Montalcini R (1956) A nerve growth-stimulating factor isolated from snake venom. Proc Natl Acad Sci USA 42: 571–574

Cohen P, Sutter A, Landreth G, Zimmermann A, Shooter EM (1980) Oxidation of tryptophan-21 alters the biological activity and receptor binding characteristics of mouse nerve growth factor. J Biol Chem 255: 2949–2954

Cole MS, Glenn JK, Coleman PD, Fahnestock MK (1986) Presence of β-nerve growth factor mRNA in human cortex from both control and Alzheimer's disease patients. Soc Neurosci Abstr 12: 1398

Darling TLJ, Fahnestock M (1988) The high molecular weight nerve growth factor complex from *Mastomys natalensis* differs from the murine nerve growth factor complex. Biochemistry 27: 6686–6692

Darling TLJ, Petrides PE, Beguin P, Frey P, Shooter EM, Selby M, Rutter WJ (1983) The biosynthesis and processing of proteins in the mouse 7S nerve growth factor complex. Cold Spring Harbor Symp Quant Biol 48: 427–434

Dicou E, Lee J, Brachet P (1986) Synthesis of nerve growth factor mRNA and precursor protein in the thyroid and parathyroid glands of the rat. Proc Natl Acad Sci USA 83: 7084–7088

Dunbar JC, Bradshaw RA (1985) Nerve growth factor biosynthesis: isolation and characterization of a guinea pig prostate kallikrein. J Cell Biochem 29: 309–319

Ebendal T, Larhammar D, Persson H (1986) Structure and expression of the chicken β nerve growth factor gene. EMBO J 5: 1483–1487

Edwards RH, Selby MJ, Rutter WJ (1986) Differential RNA splicing predicts two distinct nerve growth factor precursors. Nature 319: 784–787

Edwards RH, Selby MJ, Garcia PD, Rutter WJ (1988a) Processing of the native nerve growth factor precursor to form biologically active nerve growth factor. J Biol Chem 263: 6810–6815

Edwards RH, Selby MJ, Mobley WC, Weinrich SL, Hurby DE, Rutter WJ (1988b) Processing and secretion of nerve growth factor: expression in mammalian cells with a vaccinia virus vector. Mol Cell Biol 8: 2456–2464

Evans BA, Richards RI (1985) Genes for the α and γ subunits of mouse nerve growth factor are contiguous. EMBO J 4: 133–138

Evans BA, Drinkwater CC, Richards RI (1987) Mouse glandular kallikrein genes: structure and partial sequence analysis of the kallikrein gene locus. J Biol Chem 262: 8027–8034

Fahnestock M, Bell RA (1988) Molecular cloning of a cDNA encoding the nerve growth factor precursor from *Mastomys natalensis*. Gene 69: 257–264

Fahnestock M, Woo J, Snow J, Walz D, Mobley WC (1988) β-NGF-endopeptidase is identical to EGF-BP Type A. J Cell Biochem 12B: 278

Francke U, de Martinville B, Coussens L, Ullrich A (1983) The human gene for the β subunit of nerve growth factor is located on the proximal short arm of chromosome 1. Science 222: 1248–1251

Frazier WA, Boyd LF, Bradshaw RA (1973a) Interaction of nerve growth factor with surface membranes: biological competence of insolubilized nerve growth factor. Proc Natl Acad Sci USA 70: 2931–2935

Frazier WA, Hogue-Angeletti RA, Sherman R, Bradshaw RA (1973b) Topography of mouse 2.5S nerve growth factor. Reactivity of tyrosine and tryptophan. Biochemistry 12: 3281–3293

Furukawa Y, Furukawa S, Satoyoshi E, Hayashi K (1984) Nerve growth factor secreted by mouse heart cells in culture. J Biol Chem 259: 1259–1264

Goedert M, Fine A, Hunt SP, Ullrich A (1986) Nerve growth factor mRNA in peripheral and central rat tissues and in the human central nervous system: lesion effects in the rat brain and levels in Alzheimer's disease. Mol Brain Res 1: 85–92

Goldstein LD, Reynolds CP, Perez-Polo JR (1978) Isolation of human nerve growth factor from placental tissue. Neurochem Res 3: 175–183

Goldstein ME, Tank AW, Fossom LH, Fahnestock M, Hamill RW (1989) Androgen control of tyrosine hydroxylase and nerve growth factor in the peripheral sympathetic nervous system. Soc Neurosci Abstr 15: 1372

Greene LA, Shooter EM (1980) The nerve growth factor: biochemistry, synthesis, and mechanism of action. Annu Rev Neurosci 3: 353–402

Greene LA, Shooter EM, Varon S (1968) Enzymatic activities of mouse nerve growth factor and its subunits. Proc Natl Acad Sci USA 60: 1383–1388

Greene LA, Shooter EM, Varon S (1969) Subunit interaction and enzymatic activity of mouse 7S nerve growth factor. Biochemistry 8: 3735–3741

Harper GP, Thoenen H (1981) Target cells, biological effects, and mechanism of action of nerve growth factor and its antibodies. Annu Rev Pharmacol Toxicol 21: 205–229

Harper GP, Glanville RW, Thoenen H (1982) The purification of nerve growth factor from bovine seminal plasma. Biochemical characterization and partial amino acid sequence. J Biol Chem 257: 8541–8548

Harris-Warrick RM, Bothwell MA, Shooter EM (1980) Subunit interactions inhibit the binding of β nerve growth factor to receptors on embryonic chick sensory neurons. J Biol Chem 255: 11284–11289

Heumann R, Korsching S, Scott J, Thoenen H (1984) Relationship between levels of nerve growth factor (NGF) and its messenger RNA in sympathetic ganglia and peripheral target tissues. EMBO J 3: 3183–3189

Heumann R, Lindholm D, Bandtlow C, Meyer M, Radeke MJ, Misko TP, Shooter EM, Thoenen H (1987) Differential regulation of mRNA encoding nerve growth factor and its receptor in rat sciatic nerve during development, degeneration, and regeneration: role of macrophages. Proc Natl Acad Sci USA 84: 8735–8739

Hogue-Angeletti R, Bradshaw RA, Marshall GR (1974) Synthesis and characterization of the amino-terminal octapeptide of mouse nerve growth factor. Int J Pept Protein Res 6: 321–328

Hogue-Angeletti RA, Frazier WA, Jacobs JW, Niall HD, Bradshaw RA (1976) Purification, characterization and partial amino acid sequence of nerve growth factor from cobra venom. Biochemistry 15: 26–34

Hoffman H-D, Unsicker K (1982) The seminal vesicle of the bull: a new and very rich source of nerve growth factor. Eur J Biochem 128: 421–426

Isackson PJ, Bradshaw RA (1984) The α-subunit of mouse 7S nerve growth factor is an inactive serine protease. J Biol Chem 259: 5380–5383

Isackson PJ, Ullrich A, Bradshaw RA (1984) Mouse 7S nerve growth factor: complete sequence of a cDNA coding for the α-subunit precursor and its relationship to serine proteases. Biochemistry 23: 5997–6002

Isackson PJ, Nisco SJ, Bradshaw RA (1987) Expression of the α subunit of 7S nerve growth factor in the mouse submandibular gland. Neurochem Res 12: 959–966

Ishii DN, Shooter EM (1975) Regulation of nerve growth factor synthesis in mouse submaxillary glands by testosterone. J Neurochem 25: 843–851

Jongstra-Bilen J, Coblentz L, Shooter EM (1989) The in vitro processing of the NGF precursors by the γ-subunit of the 7S NGF complex. Mol Brain Res 5: 159–169

Katoh-Semba R, Semba R, Kashiwamata S, Kato K (1989) Sex-dependent and sex-independent distribution of the β-subunit of nerve growth factor in the central nervous and peripheral tissues of mice. J Neurochem 52: 1559–1565

Kim IS, Pantazis NJ (1985) Biological and biochemical properties of nerve growth factor synthesized by mouse S-180 cells in culture. Exp Cell Res 156: 391–404

Koo PH, Stach RW (1989) Interaction of nerve growth factor with murine α-macroglobulin. J Neurosci Res 22: 247–261

Korsching S, Thoenen H (1983) Nerve growth factor in sympathetic ganglia and corresponding target organs of the rat: correlation with density of sympathetic innervation. Proc Natl Acad Sci USA 80: 3513–3516

Korsching S, Thoenen H (1988) Developmental changes of nerve growth factor levels in sympathetic ganglia and their target organs. Dev Biol 126: 40–46

Kyte J, Doolittle RF (1982) A simple method for displaying the hydropathic character of a protein. J Mol Biol 157: 105–132

Large TH, Bodary SC, Clegg DO, Weskamp G, Otten U, Reichardt LF (1986) Nerve growth factor gene expression in the developing rat brain. Science 234: 352–355

Lärkfors L, Ebendal T (1987) Highly sensitive enzyme immunoassays for β-nerve growth factor. J Immunol Methods 97: 41–47

Levi-Montalcini R, Hamburger V (1953) A diffusible agent of mouse sarcoma producing hyperplasia of sympathetic ganglia and hyperneurotization of viscera in the chick embryo. J Exp Zool 123: 233–288

Levine JD, Taiwo YO, Woo JE, Mobley WC (1988) Hyperalgesic properties of NGF-octapeptide in the hairy skin of the rat. Soc Neurosci Abstr 14: 686

Meier R, Becker-Andre M, Gotz R, Heumann R, Shaw A, Thoenen H (1986) Molecular cloning of bovine and chick nerve growth factor (NGF): delineation of conserved and unconserved domains and their relationship to the biological activity and antigenicity of NGF. EMBO J 5: 1489–1493

Millaruelo AI, Nieto-Sampedro M, Yu J, Cotman CW (1986) Neurotrophic activity in the central and peripheral nervous system of the rat. Brain Res 374: 12–20

Mobley WC, Schenker A, Shooter EM (1976) Characterization and isolation of proteolytically modified nerve growth factor. Biochemistry 15: 5543–5552

Moore B Jr, Mobley WC, Shooter EM (1974) Proteolytic modification of the β-nerve growth factor protein. Biochemistry 13: 833–840

Mowry MD, Jensen RJ, Pantazis NJ (1984) Immunocytochemical localization and concentrations of the α and γ subunits of 7S-nerve growth factor in the submandibular gland of the mouse. Cell Tissue Res 283: 627–633

Murphy RA, Oger J, Saide JD, Blanchard MH, Arnason BGW, Hogan C, Pantazis NJ, Young M (1977a) Secretion of nerve growth factor by central nervous system glioma cells in culture. J Cell Biol 72: 769–773

Murphy RA, Singer RH, Saide JD, Pantazis NJ, Blanchard MH, Byron KS, Arnason BGW, Young M (1977b) Synthesis and secretion of a high molecular weight form of nerve growth factor by skeletal muscle cells in culture. Proc Natl Acad Sci USA 74: 4496–4500

Murphy RA, Landis SC, Bernanke J, Siminoski K (1986) Absence of α and γ subunits of 7S nerve growth factor in denervated rodent iris: immunocytochemical studies. Dev Biol 114: 369–380

Murphy RA, Chlumecky V, Smillie LB, Carpenter M, Nattriss M, Anderson JK, Rhodes JA, Barker PA, Siminoski K, Campenot RB, Haskins J (1989) Isolation and characterization of a glycosylated form of beta nerve growth factor in mouse submandibular glands. J Biol Chem 264: 12502–12509

Nichols RA, Shooter EM (1983) Characterization of the differential interaction of the microheterogeneous forms of the γ subunit of 7S nerve growth factor with natural and synthetic ligands. J Biol Chem 258: 10296–10303

Nichols RA, Shooter EM (1985) Subunit interactions of the nerve and epidermal growth factor complexes: protection of the biological subunit from proteolytic modification. Dev Neurosci 7: 216–229

Norrgren G, Ebendal T, Belew M, Jacobson C-O, Porath J (1980) Release of nerve growth factor by human glial cells in culture. Exp Cell Res 130: 31–39

Palmer TE, Neet KE (1980) Subunit interactions in 7S nerve growth factor. γ-Esterase activity as a probe of the dissociation of the 7S oligomer promoted by salt and EDTA. J Biol Chem 255: 5170–5176

Pantazis NJ (1983) Nerve growth factor synthesized by mouse fibroblast cells in culture: absence of α and γ subunits. Biochemistry 22: 4264–4271

Pantazis NJ, Jensen R (1985) The α and γ subunits of 7S nerve growth factor are present in excess concentrations in the mouse submandibular gland. Mol Cell Biochem 69: 35–48

Pantazis NJ, Blanchard MH, Arnason BGW, Young M (1977) Molecular properties of the nerve growth factor secreted by L cells. Proc Natl Acad Sci USA 74: 1492–1496

Pattison SE, Dunn MF (1975) On the relationship of zinc ion to the structure and function of the 7S nerve growth factor protein. Biochemistry 14: 2733–2739

Perez-Polo JR, Bomar H, Beck C, Hall K (1978) Nerve growth factor from *Crotalus adamenteus* snake venom. J Biol Chem 253: 6140–6148

Perez-Polo JR, Beck C, Reynolds CP, Blum M (1983) Human nerve growth factor: comparative aspects. In: Guroff G (ed) Growth and maturation factors, vol 1. Wiley, New York, pp 32–54

Ronne H, Anundi H, Rask L, Peterson PA (1979) Nerve growth factor binds to serum α-2-macroglobulin. Biophys Biochem Res Commun 87: 330–336

Rubin JS, Bradshaw RA (1981) Isolation and partial amino acid sequence analysis of nerve growth factor from the guinea pig prostate. J Neurosci Res 6: 451–464

Schwab ME, Stöckel K, Thoenen H (1976) Immunocytochemical localization of nerve growth factor (NGF) in the submandibular gland of adult mice by light and electron microscopy. Cell Tissue Res 169: 289–299

Schwarz MA, Fisher D, Bradshaw RA, Isackson PJ (1989) Isolation and sequence of a cDNA clone of β-nerve growth factor from the guinea pig prostate gland. J Neurochem 52: 1203–1209
Scott J, Selby M, Urdea M, Quiroga M, Bell GI, Rutter WJ (1983) Isolation and nucleotide sequence of a cDNA encoding the precursor of mouse nerve growth factor. Nature 302: 538–540
Selby MJ, Edwards RH, Rutter WJ (1987a) Cobra nerve growth factor: structure and evolutionary comparison. J Neurosci Res 18: 293–298
Selby MJ, Edwards R, Sharp F, Rutter WJ (1987b) Mouse nerve growth factor gene: structure and expression. Mol Cell Biol 7: 3057–3064
Server AC, Shooter EM (1976) Comparison of the arginine esteropeptidases associated with the nerve and epidermal growth factors. J Biol Chem 251: 165–173
Server AC, Shooter EM (1977) Nerve growth factor. Adv Protein Chem 31: 339–409
Server AC, Herrup K, Shooter EM, Hogue-Angeletti RA, Frazier WA, Bradshaw RA (1976) Comparison of the nerve growth factor proteins from cobra venom (*Naja naja*) and mouse submaxillary gland. Biochemistry 15: 35–39
Shelton DL, Reichardt LF (1986a) Studies on the regulation of β-nerve growth factor gene expression in the rat iris: the level of mRNA-encoding nerve growth factor is increased in irises placed in explant cultures in vitro, but not in irises deprived of sensory or sympathetic innervation in vivo. J Cell Biol 102: 1940–1948
Shelton DL, Reichardt LF (1986b) Studies on the expression of the β nerve growth factor (NGF) gene in the central nervous system: level and regional distribution of NGF mRNA suggest that NGF functions as a trophic factor for several distinct populations of neurons. Proc Natl Acad Sci USA 83: 2714–2718
Siminoski K, Murphy RA (1987) Detection of molecules with nerve growth factor binding activity in medium conditioned by L-929 fibroblasts. Brain Res 435: 273–282
Smith AP, Varon S, Shooter EM (1968) Multiple forms of the nerve growth factor protein and its subunits. Biochemistry 7: 3259–3268
Stach RW, Shooter EM (1974) The biological activity of cross-linked β nerve growth factor protein. J Biol Chem 249: 6668–6674
Stach RW, Shooter EM (1980) Cross-linked 7S nerve growth factor is biologically inactive. J Neurochem 34: 1499–1505
Stach RW, Server AC, Pignatti P-F, Piltch A, Shooter EM (1976) Characterization of the γ subunits of the 7S nerve growth factor complex. Biochemistry 15: 1455–1461
Suda K, Barde Y-A, Thoenen H (1978) Nerve growth factors in mouse and rat serum: correlation between bioassay and radioimmunoassay determinations. Proc Natl Acad Sci USA 75: 4042–4046
Thoenen H, Barde Y-A (1980) Physiology of nerve growth factor. Physiol Rev 60: 1284–1335
Thomas KA, Baglan NC, Bradshaw RA (1981a) The amino acid sequence of the γ-subunit of mouse submaxillary gland 7S nerve growth factor. J Biol Chem 256: 9156–9166
Thomas KA, Silverman RE, Jeng I, Baglan NC, Bradshaw RA (1981b) Electrophoretic heterogeneity and polypeptide chain structure of the γ-subunit of mouse submaxillary 7S nerve growth factor. J Biol Chem 256: 9147–9155
Ullrich A, Gray A, Berman C, Dull TJ (1983) Human β-nerve growth factor gene sequence highly homologous to that of mouse. Nature 303: 821–825
Ullrich A, Gray A, Wood WI, Hayflick J, Seeburg P (1984) Isolation of a cDNA clone coding for the γ subunit of nerve growth factor using a high-stringency selection procedure. DNA 3: 387–392
Varon S, Shooter EM (1970) The nerve growth factor proteins of the mouse submaxillary gland. In: Bowman RE, Datta SP (eds) Biochemistry of brain and behavior. Plenum, New York, pp 41–64
Varon S, Nomura J, Shooter EM (1968) Reversible dissociation of the mouse nerve growth factor protein into different subunits. Biochemistry 7: 1296–1303
Walker P, Weichsel ME Jr, Fisher DA (1980) Human nerve growth factor: lack of immunocrossreactivity with mouse nerve growth factor. Life Sci 26: 195–200
Werrbach-Perez K, Perez-Polo JR (1987) De novo synthesis of NGF subunits in S-180 mouse sarcoma cell line. Neurochem Res 12: 875–883
Whitemore SR, Friedman PL, Larhammar D, Persson H, Gonzalez-Carvajal M, Holets VR (1988) Rat beta-nerve growth factor sequence and site of synthesis in the adult hippocampus. J Neurosci Res 20: 403–410
Wion D, Perret C, Prechin N, Keller A, Behar G, Brachet P, Auffray C (1986) Molecular cloning of the avian β-nerve growth factor gene: transcription in brain. FEBS Lett 203: 82–86

Young M, Saide JD, Murphy RA, Blanchard MH (1978) Nerve growth factor: multiple dissociation products in homogenates of the mouse submandibular gland. Purification and molecular properties of the intact undissociated form of the protein. Biochemistry 17: 1490–1498

Young M, Blanchard MH, Sessions F, Boyle MDP (1988) Subunit structure of high molecular weight mouse nerve growth factor. Biochemistry 27: 6675–6681

Zabel BU, Eddy RL, Lalley PA, Scott J, Bell GI, Shows TB (1985) Chromosomal locations of the human and mouse genes for precursors of epidermal growth factor and the β subunit of nerve growth factor. Proc Natl Acad Sci USA 82: 469–473

Zheng M, Heinrich G (1988) Structural and functional analysis of the promoter region of the nerve growth factor gene. Mol Brain Res 3: 133–140

Biochemical Characterization of the Nerve Growth Factor Receptor in Neural-Related Tumors*

A. H. Ross[1]**, G. Sobue[2], H. Hotta[3], and U. R. Reddy[4]

1 Introduction

Neurotrophic factors are required for regulation of cell division, cell death, axon outgrowth and synapse formation during fetal development, maintenance of the adult nervous system, and neural regeneration following injury. Due to the relative abundance of nerve growth factor (NGF) in the murine salivary gland, NGF was the first neurotrophic factor to be isolated and characterized (Levi-Montalcini 1987).

The NGF receptor was detected by ^{125}I-NGF binding studies, but initial attempts to determine the molecular identity of the receptor were confusing and contradictory. Recently, the NGF receptor was identified using anti-receptor monoclonal antibodies (MAb), and the gene for the receptor was cloned. In this article, we review these recent advances and then focus on current problems concerning NGF receptor structure and its possible role in neural-related tumors.

* This investigation was supported by USPHS grants NS-21716 and CA-47983 and by the National Neurofibromatosis Foundation

** Corresponding author
[1] The Worcester Foundation for Experimental Biology, Shrewsbury, MA 01545, USA
[2] Division of Neurology, Aichi Medical University, Nagakute, Aichi 480-11, Japan
[3] Department of Microbiology, Kobe University School of Medicine, Kobe 650, Japan
[4] Department of Neurology, The Childrens' Hospital of Philadelphia, Philadelphia, PA 19104, USA

Current Topics in Microbiology and Immunology, Vol. 165

2 NGF

NGF is a 26 kDa polypeptide neurotrophic factor capable of mediating a variety of biological responses (LEVI-MONTALCINI 1987). Serum levels are quite low (STEPHANI et al. 1987) suggesting that this factor is synthesized at its site of action. It acts as a survival factor for sympathetic and sensory neurons both in vivo (JOHNSON et al. 1980) and in culture (HERRUP and SHOOTER 1975) and also can act as a chemoattractant (LEVI-MONTALCINI 1976). In addition, NGF is a differentiation factor for the pheochromocytoma cell line, PC12 (GREENE and TISCHLER 1976), neuroblastomas (SONNENFELD and ISHII 1982), fetal peripheral neurons (COLLINS and DAWSON 1983), fetal brain cholinergic neurons (HEFTI et al. 1986), and fetal adrenal chromaffin cells (LILLIEN and CLAUDE 1985) and may have similar effects in vivo (LEVI-MONTALCINI and ALOE 1985; GOEDERT et al. 1984). The innervated organs provide a source of NGF for NGF dependent neurons, and during development competition for NGF may regulate neuronal cell death (PURVES and LICHTMAN 1985). Although NGF was originally named a growth factor, its role as a mitogen is very limited, and there is no evidence that NGF acts in vivo as a mitogen. PC12 cells cease to divide after exposure to NGF, although a variant of PC12 cells (BURSTEIN and GREENE 1982) grows continuously in response to NGF. Fetal chromaffin cells (LILLIEN and CLAUDE 1985) divide several times in response to NGF before terminally differentiating. Some neuroblastoma cell lines differentiate in response to NGF, but there is disagreement whether cell division stops (PEREZ-POLO et al. 1979; SONNENFELD and ISHII 1982).

3 Binding Studies with NGF

Binding studies with ^{125}I-NGF have demonstrated the presence of specific receptors on the surface of a variety of cell types, particularly those of neural crest origin (VALE and SHOOTER 1985). Such receptors are found on all NGF-responsive cells, such as sensory and sympathetic neurons, neuroblastoma cell lines, and the rat PC12 cell line (Table 1). Detailed analysis of ^{125}I-NGF binding generally shows that receptors are heterogeneous with regard to NGF binding properties, and most studies have concluded that there are two forms of the NGF receptor, differing in their affinity for NGF and in the rate with which they bind and release NGF. The low- and high-affinity receptors bind NGF with a K_d of about 2 nM and 0.2 nM, respectively, and the difference in affinity is primarily due to the slower off-rate of the high-affinity receptor. Also, the two receptor types differ in their sensitivity to proteases and in their solubility in nonionic detergents (SCHECHTER and BOTHWELL 1981; VALE et al. 1985). The high-affinity NGF receptor appears to be largely responsible for NGF internalization and for the

Table 1. Low- and high-affinity NGF receptors on normal and transformed cells

Cell type	Receptor sites per cell		Reference(s)
	Low affinity	High affinity	
Dorsal root ganglia neurons	45000	3000	SUTTEER et al. (1979)
Sympathetic neurons	27000	2100	GODFREY and SHOOTER (1986)
Schwann cells	33000	0	YASUDA et al. (1987), DISTEFANO and JOHNSON (1988)
Neurofibroma cells	70000	0	SONNENFELD et al. (1986)
PC12 cells	50000	2500	SCHECHTER and BOTHWELL (1981)
Neuroblastoma SH-SY5Y	0	700	SONNENFELD and ISHII (1985)
Neuroblastoma MC-IXC	250000	0	SONNENFELD and ISHII (1985)

differentiating response in PC12 cells (GREEN et al. 1986). As judged by dose-response curves, NGF-induced modulation of PC12 tyrosine hydroxylase and amino acid transport activities are mediated by both low- and high-affinity NGF receptors (KEDES et al. 1982; ROWLAND et al. 1987). For bovine chromaffin cells which lack high-affinity sites, tyrosine hydroxylase activity is modulated by low-affinity sites (ACHESON et al. 1984). Schwann cells which have only low-affinity NGF receptors display altered levels of the cell-cell adhesion molecule L1 following exposure to NGF (SEILHEIMER and SCHACHNER 1987).

4 Biochemical Characterization of the NGF Receptor

Initial biochemical characterization of NGF receptors led to a confusing variety of putative receptor forms. Hydrodynamic studies of the detergent-solubilized receptor of rabbit superior cervical ganglia revealed a highly asymmetric 135 kDa molecule (COSTRINI et al. 1979). Chemical cross linking of ^{125}I-NGF to these receptors revealed forms of 130 kDa and 100 kDa (MASSAGUE et al. 1982). Similar cross-linking studies on rat PC12 pheochromocytoma cells led to receptor forms of 135–145 kDa and 107–117 kDa, (MASSAGUE et al. 1982), while cross-linking characterization of substantially purified receptor from A875 melanoma cells revealed receptors of 70–90 kDa and 180–200 kDa (GROB et al. 1983; PUMA et al. 1983). The cause of this confusion is still not clear.

A critical advance was the development of murine MAbs directed against the human melanoma NGF receptor (ROSS et al. 1984) and against the rat NGF receptor (CHANDLER et al. 1984). Intact human melanoma cells and detergent solubilized PC12 membranes were used as immunogens for the anti-human and anti-rat receptor MAbs, respectively. The former MAbs were prepared as part of a large program to develop anti-tumor MAbs and were not initially recognized as anti-NGF receptor MAbs (HERLYN et al. 1983). We focused on these MAbs

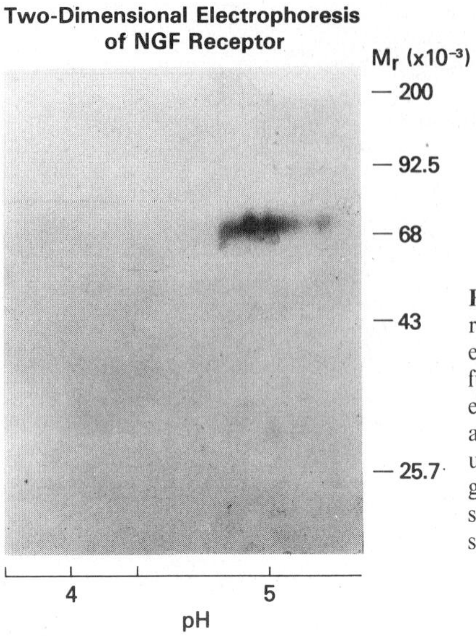

Fig. 1. Two-dimensional gel electrophoresis of NGF receptor. A875 cells were labeled with [^{35}S]cysteine, extracted with detergent, and clarified by ultracentrifugation. Immunoprecipitates were prepared using either ME20.4 or P3X63Ag8 (control) IgG and analyzed by two-dimensional gel electrophoresis using ampholytes in the pH3-5 range. An autoradiogram of the ME20.4 gel is shown. There were no spots evident on the P3X63Ag8 autoradiogram (not shown)

because of their preferential binding to cell lines rich in NGF receptor and determined that these MAbs block binding of ^{125}I-NGF to melanoma cells and immunoprecipitate ^{125}I-NGF–NGF receptor complexes prepared by chemical cross-linking with ethyldimethylisopropylaminocarbodiimide (Ross et al. 1984). The anti-rat receptor MAb also immunoprecipitates the ^{125}I-NGF–NGF receptor complex but does not inhibit binding of ^{125}I-NGF to PC12 cells. However, the MAb does inhibit NGF-induced neurite formation by PC12 cells (Chandler et al. 1984). Also, this MAb is retrogradely transported by nerve axons in a manner similar to that known for NGF (Taniuchi and Johnson 1985). Both MAbs are highly species-specific. The MAbs prepared against human cells bind to human or monkey cells (Marano et al. 1987) but not to lower species. MAb 192 binds to rat NGF receptor but not to chick, mouse, or human receptors.

Initial studies revealed that both MAbs react to the same antigen, a 75 kDa protein as determined by SDS-PAGE (Grob et al. 1985; Taniuchi et al. 1986). Two-dimensional gel electrophoresis of [^{35}S] cysteine labeled receptor revealed that the receptor was a heterogeneous, acidic protein (Fig. 1) suggesting post translational modifications of the receptor. We have identified at least two and probably three such modifications. Pulse-chase analysis demonstrated that the receptor is initially synthesized as a 62 kDa protein and undergoes a tunicamycin sensitive modification within the first 30–60 min following synthesis. This step is clearly N-linked glycosylation. There follows a slower tunicamycin insensitive step which is probably O-linked glycosylation. Sequencing of the corresponding cDNA (see Chapt. 4) revealed a serine, threonine rich region which is characteristic of O-linked glycosylation sites. However, there is no direct chemical evidence that there really is O-linked glycosylation.

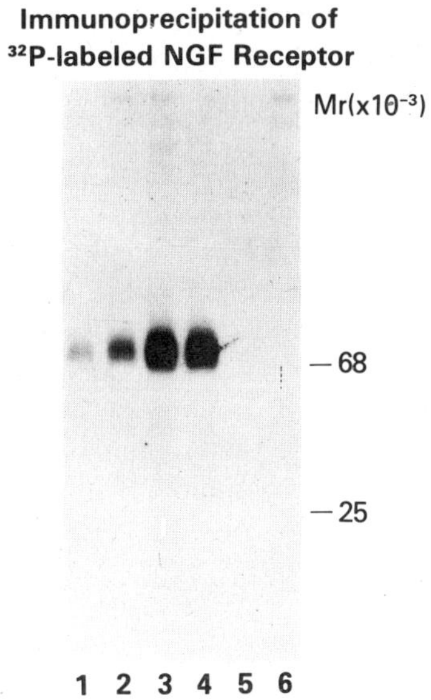

Fig. 2. Immunoprecipitation of [^{32}P]phosphate labeled NGF receptor. A875 cells in the absence (*lanes 1, 2* and *5*) or presence (*lanes 2, 4* and *6*) of 50 ng/ml NGF were labeled with [^{32}P]phosphate, solubilized with detergent, clarified by ultracentrifugation, and used for immunoprecipitation using anti-NGF receptor MAb ME82-11 (*lanes 1* and *2*), anti-NGF receptor MAb ME20.4 (*lanes 3* and *4*), or control MAb P3X63Ag8 (*lanes 5* and *6*). The resulting samples were boiled with SDS sample buffer containing 2-mercaptoethanol and analyzed by SDS-polyacrylamide (10%) gel electrophoresis and autoradiography

In both human (GROB et al. 1985) and rat cells (TANIUCHI et al. 1986), the receptor is phosphorylated (Fig. 2). The receptor is mainly phosphorylated on serine residues but there is also a small amount of phosphorylation of threonine residues. Peptide mapping revealed that the phosphorylation is confined to two tryptic peptides. The more hydrophobic peptide accounted for most of the phosphoserine, and the more hydrophilic peptide had small amounts of both phosphoserine and phosphothreonine (TANIUCHI et al. 1986). Neither the identity of the kinase responsible for phosphorylation of the receptor nor the function of the phosphorylation is known. The extent of phosphorylation is not effected by NGF treatment or by phorbol esters, and the isolated NGF receptor has no phosphorylation activity (TANIUCHI et al. 1986; MARANO et al. 1987). Since protein phosphorylation is usually regulatory the exact function of this modification will probably have to await a better understanding of the mechanism by which NGF receptor triggers neuronal differentiation.

Analysis of NGF receptor by SDS-PAGE results in at least two bands (GROB et al. 1985). The major bands are 75 and 200 kDa, but there are sometimes minor bands of even higher apparent molecular mass. Utilizing two-dimensional SDS-PAGE, in which the first dimension was nonreducing and the second reducing, it was demonstrated that the higher molecular mass bands are multimers of the 75 kDa protein band. It is not known what the state of aggregation of the receptor is in intact biological membranes. This is an important question since ligand-induced aggregation has been shown to be an important step in the response of other cell surface receptors (SCHLESSINGER 1986; METZGER et al. 1986).

To allow a more detailed chemical study, the NGF receptor was isolated from cultured human melanoma cells by a combination of lectin affinity chromatography and immunoaffinity chromatography (MARANO et al. 1987). A single 75 kDa protein species was detected which bound the anti-NGF receptor MAbs and NGF. Hence, this was the correct protein, and no other protein component is required for specific NGF binding. The isolated receptor protein was used for two applications. First, this material was used as an efficient immunogen to generate more anti-receptor MAbs (MARANO et al. 1987). Many of these new MAbs have properties similar to the original MAbs suggesting that there may be an immunodominant epitope. Second, the isolated receptor was subjected to microsequencing of the N-terminus (MARANO et al. 1987). This sequence is very similar to the corresponding sequence generated from the rat NGF receptor and was useful for cloning of the NGF receptor gene (CHAO et al. 1986; JOHNSON et al. 1986; RADECKE et al. 1987).

5 Future Directions

The identity of the high-affinity NGF receptor is a central focus of the field. HOSANG and SHOOTER (1987) reported that the high-affinity receptor was a 140 kDa protein distinct from the 75 kDa receptor, but GREEN and GREENE (1986) could not reproducibly detect the second protein and presented strong evidence that the 75 kDa protein (gp75) was a component of both the low- and high-affinity receptors. These data are reinforced by the finding that the anti-rat NGF receptor MAb inhibits NGF-induced neurite extension (CHANDLER et al. 1984) and that gp75 on some neuroblastoma cells is high affinity (CHAO et al. 1986). Also, anti-gp75 antibodies prepared against synthetic peptides inhibit neurite extension by PC12 cells (EVELETH and BRADSHAW 1988).

Given the strong evidence for a role of gp75 in the high-affinity binding site, the question becomes what modification of gp75 results in high-affinity binding. There are three models currently under consideration. First, gp75 might undergo a conformational change or a change in lateral association (PUMA et al. 1983) perhaps analagous to the insulin-like growth factor I receptor for which the receptor monomer binds insulin-like growth factor I with low affinity and the receptor dimer binds insulin-like growth factor I with high affinity (TOLLEFSEN and THOMPSON 1988). There is surprisingly little data concerning the state of aggregation of membrane-bound NGF receptor although detergent solubilized receptors tend to form a mixture of monomers and dimers (GROB et al. 1985). Second, there might be a second NGF binding protein in cells with high-affinity receptors. The high-affinity receptor is proposed to be a heterodimer of the two NGF binding proteins. This model gained great popularity following the discovery of a second binding protein for interleukin-2 (TESHIGAWARA et al. 1987), but now after efforts by many laboratories the second NGF binding protein has not been reproducibly detected. In the third model, the high-affinity receptor is proposed to be a heterodimer of gp75 with a modulator protein which does not bind NGF but rather enhances the affinity of gp75 for NGF. Such a modulator protein might be difficult to detect particularly if the association occurs between transmembrane domains or intracellular domains. For the IgE receptor, the IgE binding subunit was easily detected, but other components of the receptor were discovered only after years of additional study (METZGER et al. 1986).

The mechanism of NGF receptor action is poorly understood. There are many biochemical changes immediately following NGF stimulation (YANKNER and SHOOTER 1982), but the primary action of the receptor is not known. The receptor is not a protein kinase, and the sequence of the receptor is not significantly homologous to any known enzyme. Recently, MAHER (1988) reported that NGF treatment of PC12 cells induced phosphotyrosine-bearing proteins. These results are very important since a connection between the receptor and a tyrosine-specific kinase might allow the first biochemical delineation of the mechanism of NGF receptor action.

6 Expression of NGF Receptor by Neural-Related Tumors

Due to the paucity of material available from primary neural cultures, virtually all of the biochemical studies of the NGF receptor have utilized neural-related tumors. A wide variety of neural-related tumors express NGF receptor (Table 2). Although melanomas do not display a neuronal phenotype they are of neural

Table 2. Human neural-related tumors expressing NGF receptor

Cell type	Sample	Reference(s)
Melanoma	Cell lines Tumor sections	FABRICANT et al. (1977), HERLYN et al. (1983), ROSS et al. (1984), CHESA et al. (1988)
Neuroblastoma	Cell lines	SONNENFELD and ISHII (1985), RETTIG et al. (1987)
Ganglioneuroma	Tumor sections	CHESA et al. (1988)
Meningioma	Tumor sections	CHESA et al. (1988)
Pheochromocytoma	Tumor sections	ROSS et al. (1984)
Ewing's sarcoma	Cell lines Tumor sections	LIPINSKI et al. (1987), CHESA et al. (1988)
Embryonal rhabdomyosarcoma	Tumor sections	CHESA et al. (1988)
Synovial sarcoma	Tumor sections	CHESA et al. (1988)
Neurofibroma	Primary cultures Tumor sections	ROSS et al. (1986)
Schwannoma	Tumor sections	CHESA et al. (1988)
Occasional carcinomas and lymphomas	Tumor sections	CHESA et al. (1988)

crest origin. Melanoma cell lines have been of particular utility because some of these lines express unusually large numbers of NGF receptors (Table 1). However, there is only a slight biological response by these cell lines to NGF (FABRICANT et al. 1977). Neuroblastoma cell lines vary in NGF receptor expression. Neuroepithelioma cell line MCIXC expresses large numbers of receptors but is apparently nonresponsive (SONNENFELD and ISHII 1985). Neuroblastoma cell line SY5Y has very few receptors, but these are high-affinity sites. This line responds to NGF by neurite extension but does not cease cell division (SONNENFELD and ISHII 1982). Many freshly excised neuroblastomas apparently lack NGF receptor but the more differentiated ganglioneuromas are receptor positive (CHESA et al. 1988). Short-term cultures of human pheochromocytomas differentiate and cease cell division in response to NGF (TISCHLER et al. 1980). Ewing's sarcoma was assigned as a neural-derived tumor based on expression of neural markers such as NGF receptor (LIPINSKI et al. 1987), but there is not report of a Ewing's sarcoma biological response to NGF. Two other small round cell tumors, embryonal and synovial rhabdomyosarcomas, are also positive (CHESA et al. 1988). Neurofibromas express the low-affinity NGF receptor and do not have any apparent response to NGF (PLEASURE et al. 1986; ROSS et al. 1986; SONNENFELD et al. 1986); Schwannomas are also positive. In unpublished studies, we have demonstrated that medulloblastoma cell lines express NGF receptor (BAKER et al., unpublished work). The affinity of the medulloblastoma NGF receptor and the response of these cells to NGF is not known.

The origin of the widespread expression of NGF receptor by neural-related tumors is not clear. Overexpression of the epidermal growth factor receptor frequently involves amplification of the receptor gene. We have screened a series of melanoma cell lines which strongly express NGF receptor at the protein and RNA levels. Southern blot analysis show no amplification or apparent modification of the NGF receptor gene (Fig. 3). The regulation of NGF receptor expression may be quite complex. The promotor is apparently constituitively active and regulation is due to other genetic elements (SEHGAL et al. 1988). Frequently, tumors express antigens associated with fetal development, i.e. oncofetal antigens. Since NGF receptor is expressed most strongly during fetal development (YAN and JOHNSON 1987; SCARPINI et al. 1988), it might be expressed simply as part of the shift to the oncofetal program of gene expression. It is intriguing to consider a causal role for NGF receptor overexpression in tumor formation, but there is little supporting evidence. NGF has been reported to induce sustained cell proliferation in only one case, a variant PC12 subline (BURSTEIN and GREENE 1982). Also, there is no known pattern of coexpression of NGF and NGF receptor by neural-related tumors.

In normal development, NGF induces terminal differentiation, and lack of NGF induces cell death. Hence, a more reasonable hypothesis is that the activation of an oncogene retards the cell differentiation normally induced by NGF and other factors and, thereby, contributes to tumor formation. In a series of infant autopsies, a high frequency of adrenal lesions histologically identical to

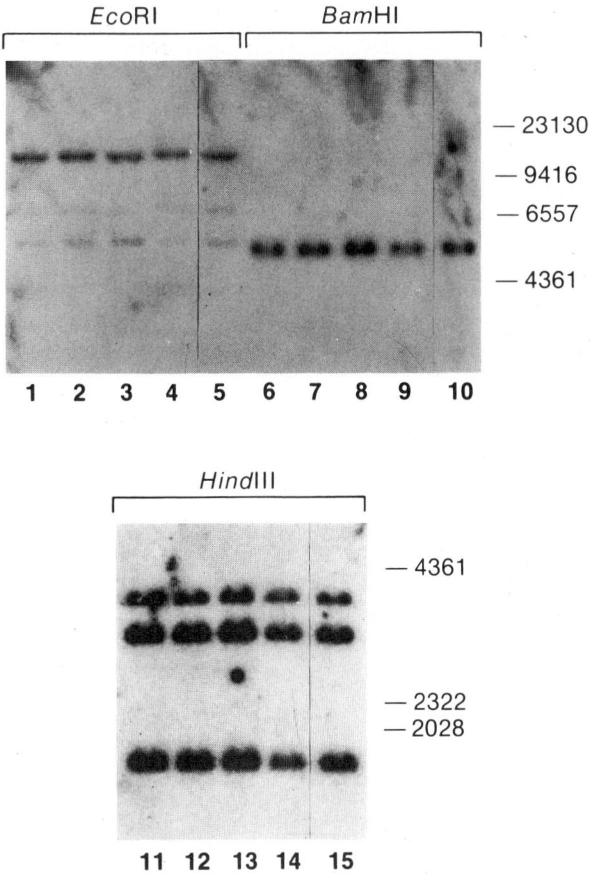

Fig. 3. Southern blotting analysis of melanoma cell lines. Total cellular DNA (10 μg per lane) was digested with *Eco*R1 (*lanes 1–5*), *Bam*H1 (*lanes 6–10*), or *Hind*III (*lanes 11–15*) and separated electrophoretically on a 0.8% agarose gel. After transfer to a nitrocellulose filter, the DNA was probed with NGF receptor cDNA plasmid pH3-1. *Lanes 1, 6,* and *11* A875 cells; *Lanes 2, 7,* and *12* HS294 cells; *Lanes 3, 8,* and *13* WM35 cells; *Lanes 4, 9,* and *14* WM9 cells; *Lanes 5, 10,* and *15* Epstein-Barr virus-transformed lymphocyte cell line derived from the same patient as melanoma cell line WM9

neuroblastomas was observed (Beckwith and Perrin 1963). In view of the known clinical incidence of neuroblastoma, the majority of these lesions must differentiate during normal growth and development. Hence, neuroblastomas may mimic a normal stage of fetal development, and factors which induce normal differentiation may have therapeutic potential. The development of such therapeutic approaches is particularly urgent for childhood tumors such as neuroblastoma. Physicians treating newborn patients are severely limited in the doses of radiotherapy and chemotherapy. NGF is a potent stimulus for cell differentiation but it is also a "sticky" molecule which may be difficult to deliver to the tumor site. For this reason, it is critical to understand the mechanism by which NGF acts and to develop new drugs to control this pathway.

References

Acheson AL, Naujoks K, Thoenen H (1984) Nerve growth factor-mediated enzymic induction in primary cultures of bovine adrenal chromaffin cells. J Neurosci 4: 1771–1780
Beckwith JB, Perrin EV (1963) In situ neuroblastomas: a contribution to the natural history of neural crest tumors. Am J Pathol 43: 1089–1104
Burstein DE, Greene LA (1982) Nerve growth factor has both mitogenic and antimitogenic activity. Dev Biol 94: 477–482
Chandler CE, Parsons LM, Hosang M, Shooter EM (1984) A monoclonal antibody modulates the interaction of nerve growth factor with PC12 cells. J Biol Chem 259: 6882–6889
Chao M, Bothwell M, Ross AH, Koprowski H, Lanahan AA, Buck CR, Sehgal A (1986) Gene transfer and molecular cloning of the human NGF receptor. Science 232: 518–521
Chesa PG, Rettig WJ, Thomson TM, Old LJ, Melamed MR (1988) Immunohistochemical analysis of nerve growth factor receptor expression in normal and malignant human tissues. J Histochem Cytochem 36: 383–389
Collins F, Dawson A (1983) An effect of nerve growth factor on parasympathetic neurite outgrowth. Proc Natl Acad Sci USA 80: 2091–2094
Costrini NV, Kogan M, Kukreja K, Bradshaw RA (1979) Physical properties of the detergent-extracted nerve growth factor receptor of sympathetic ganglia. J Biol Chem 254: 11242–11246
DiStefano PS, Johnson EM (1988) Nerve growth factor receptors on cultured rat Schwann cells. J Neurosci 8: 231–241
Eveleth DD, Bradshaw RA (1988) Antibodies directed against NGF receptor. Soc Neurosci Abstr 685
Fabricant RN, DeLarco JE, Todaro GJ (1977) Nerve growth factor receptors on human melanoma cells in culture. Proc Natl Acad Sci USA 74: 565–569
Godfrey EW, Shooter EM (1986) Nerve growth factor receptors on chick embryo sympathetic ganglion cells: binding characteristics and development. J Neurosci 6: 2543–2550
Goedert M, Otten U, Hunt SP, Bond A, Chapman D, Schlumpf M, Lichtensteiger W (1984) Biochemical and anatomical effects of antibodies against nerve growth factor on developing rat sensory ganglia. Proc Natl Acad Sci USA 81: 1580–1584
Green SH, Greene LA (1986) A single $M_r \approx 103{,}000$ ^{125}I-ß-nerve growth factor-affinity-labeled species represents both the low and high affinity forms of the nerve growth factor receptor. J Biol. Chem 261: 15316–15326
Greene LA, Tischler (1976) Establishment of a noradrenergic clonal line of rat adrenal pheochromocytoma cells which respond to nerve growth factor. Proc Natl Acad Sci USA 73: 2424–2428
Green SH, Rydel RE, Connolly JL, Greene LA (1986) PC12 mutants that possess low- but not high-affinity nerve growth factor receptors neither respond to nor internalize nerve growth factor. J Cell Biol 102: 830–843
Grob PM, Berlot CH, Bothwell MA (1983) Affinity labeling and partial purification of nerve growth factor receptors from rat pheochromocytoma and human melanoma cells. Proc Natl Acad Sci USA 80: 6819–6823
Grob PM, Ross AH, Koprowski H, Bothwell M (1985) Characterization of the human melanoma nerve growth factor receptor. J Biol Chem 260: 8044–8049
Hefti F, Hartikka J, Salvatierra A, Weiner WJ, Mash DC (1986) Localization of nerve growth factor receptors in cholinergic neurons of the human basal forebrain. Neurosci Lett 69: 37–41
Herlyn M, Steplewski Z, Herlyn D, Clark WH, Ross AH, Blaszczyk M, Pak KY, Koprowski H (1983) Production and characterization of monoclonal antibodies against human malignant melanoma. Cancer Invest 1: 215–224
Herrup K, Shooter EM (1975) Properties of the ß-nerve growth factor receptor in development. J Cell Biol 67: 118–125
Hosang M, Shooter EM (1987) The internalization of nerve growth factor high-affinity receptors on pheochromocytoma PC12 cells. EMBO J 6: 1197–1202

Johnson D, Lanahan A, Buck CR, Sehgal A, Morgan C, Mercer E, Bothwell M, Chao M (1986) Expression and structure of the human NGF receptor. Cell 47: 545–554

Johnson EM, Gorin PD, Brandies LD, Pearson J (1980) Dorsal root ganglion neurons are destroyed by exposure in utero to maternal antibody to nerve growth factor. Science 210: 916–918

Kedes DH, Gunning PW, Shooter EM (1982) Nerve growth factor-induced stimulation of aminobutyric acid uptake in PC12 cells. J Neurosci Res 8: 357–365

Levi-Montalcini R (1976) Nerve growth factor. Prog Brain Res 45: 235–256

Levi-Montalcini R (1987) The nerve growth factor 35 years later. Science 237: 1154–1162

Levi-Montalcini R, Aloe L (1985) Differentiating effects of murine nerve growth factor in the peripheral and central nervous systems of *Xenopus laevis* tadpoles. Proc Natl Acad Sci USA 82: 7111–7115

Lillien LE, Claude P (1985) Nerve growth factor is a mitogen for cultured chromaffin cells. Nature 317: 632–634

Lipinski M, Braham K, Philip I, Wiels J, Philip T, Goridis C, Lenoir GM, Tursz T (1987) Neuroectoderm-associated antigens on Ewing's sarcoma cell lines. Cancer Res 47: 183–187

Maher PA (1988) Nerve growth factor induces protein-tyrosine phosphorylation. Proc Natl Acad Sci USA 85: 6788–6791

Marano N, Dietzschold B, Earley JJ, Schatteman, G, Thompson S, Grob P, Ross AH, Bothwell M, Atkinson BF, Koprowski H (1987) Purification and amino terminal sequencing of human melanoma nerve growth factor receptor. J Neurochem 48: 225–232

Massague, J, Bixser S, Johnsor GL, Czech MP (1982) Affinity labeling of a nerve growth factor receptor component on rat pheochromocytoma (PC12) cells. Biochim Biophys Acta 693: 205–212

Metzger H, Alcaraz G, Hohman R, Kinet J-P, Pribluda V, Quarto R (1986) The receptor with high affinity for immunoglobulin E. Annu Rev Immunol 4: 419–470

Perez-Polo JR, Werrbach-Perez K, Tiffany-Castiglioni E (1979) A human clonal cell line model of differentiating neurons. Dev Biol 71: 341–355

Pleasure D, Kreider B, Sobue G, Ross AH, Koprowski H, Sonnenfeld KH, Rubenstein AE (1986) Schwann-like cells cultured from human dermal neurofibromas. Ann NY Acad Sci 486: 227–240

Puma P, Buxser SE, Watson L, Kelleher J, Johnson GL (1983) Purification of the receptor for nerve growth factor from A875 melanoma cells by affinity chromatography. J Biol Chem 258: 3370–3375

Purves D, Lichtman JW (1985) Principles of neural development. Sinauer Assoc, Sunderland UK

Radeke MJ, Misko TP, Hsu C, Herzenberg LA, Shooter EM (1987) Gene transfer and molecular cloning of the rat nerve growth factor receptor. Nature 325: 593–597

Rettig WJ, Spengler BA, Chesa PG, Old LJ, Biedler JL (1987) Coordinate changes in neuronal phenotype and surface antigen expression in human neuroblastoma cell variants. Cancer Res 47: 1383–1389

Ross AH, Grob P, Bothwell M, Elder DE, Ernst CS, Marano N, Ghrist BFD, Slemp CC, Herlyn M, Atkinson B, and Koprowski H (1984) Characterization of nerve growth factor receptor in neural crest tumors using monoclonal antibodies. Proc Natl Acad Sci USA 81: 6681–6685

Ross AH, Pleasure D, Sonnenfeld K, Atkinson B, Kreider B, Jackson DM, Taff I, Scarpini E, Lisak RP, Koprowski H (1986) Expression of melanoma-associated antigens by normal and neurofibroma Schwann cells. Cancer Res 46: 5887–5892

Rowland EA, Muller TH, Goldstein M, Greene LA (1987) Cell-free detection and characterization of a novel nerve growth factor-activated protein kinase in PC12 cells. J Biol Chem 262: 7504–7513

Scarpini E, Ross AH, Rosen JL, Brown MJ, Rostami A, Koprowski H, Lisak RP (1988) Expression of nerve growth factor receptor during human peripheral nerve development. Dev Biol 125: 301–310

Schechter AL, Bothwell MA (1981) Nerve growth factor receptors on PC12 cells: evidence for two receptor classes with differing cytoskeletal association. Cell 24: 867–874

Schlessinger J (1986) Allosteric regulation of the epidermal growth factor receptor kinase. J Cell Biol 103: 2067–2072

Sehgal A, Patil N, Chao M (1988) A constituitive promoter directs expression of the nerve growth factor gene. Mol Cell Biol 8: 3160–3167

Seilheimer B, Schachner M (1987) Regulation of neural cell adhesion molecule expression on cultured mouse Schwann cells by nerve growth factor. EMBO J 6: 1611–1616

Sonnenfeld KH, Ishii DN (1982) Nerve growth factor effects and receptors in cultured human neuroblastoma cell lines. J Neurosci Res 8: 375–391

Sonnenfeld KH, Ishii DN (1985) Fast and slow nerve growth factor binding sites in human neuroblastoma and rat pheochromocytoma cell lines. J Neurosci 5: 1717–1728

Sonnenfeld KH, Bernd P, Sobue G, Lebwohl M, Rubenstein AE (1986) Nerve growth factor receptors on dissociated neurofibroma Schwann-like cells. Cancer Res 46: 1446–1452
Stephani U, Sutter A, Zimmermann A (1987) Nerve growth factor in serum. J Neurosci Res 17: 25–35
Sutter A, Riopelle RJ, Harris-Warrick RM, Shooter EM (1979) Nerve growth factor receptors. J Biol Chem 254: 5972–5982
Taniuchi M, Johnson EM (1985) Characterization of the binding properties and retrograde transport of a monoclonal antibody directed against the rat nerve growth factor receptor. J Cell Biol 101: 1100–1106
Taniuchi M, Johnson EM, Roach PJ, Lawrence JC (1986) Phosphorylation of nerve growth factor receptor proteins in sympathetic neurons and PC12 cells. J Biol Chem 261: 13342–13349
Teshigawara K, Wang H-M, Kato K, Smith KA (1987) Interleukin 2 high-affinity receptor expression requires two distinct binding proteins. J Exp Med 165: 223–238
Tischler AS, Delellis RA, Biales B, Nunnemacher G, Carabba V, Wolfe HJ (1980) Nerve growth factor-induced neurite outgrowth from normal human chromaffin cells. Lab Invest 43: 399–409
Tollefsen SE, Thompson K (1988) The structural basis for insulin-like growth factor I receptor high affinity binding. J Biol Chem 263: 16267–16273
Vale RD, Shooter EM (1985) Assaying binding of nerve growth factor to cell surface receptors. Methods Enzymol 109: 21–39
Vale RD, Ignatius MJ, Shooter EM (1985) Association of nerve growth factor receptors with the Triton X-100 cytoskeleton of PC12 cells. J Neurosci 5: 2762–2770
Yan Q, Johnson EM (1987) A quantitative study of the developmental expression of nerve growth factor (NGF) receptor in rats. Dev Biol 121: 139–148
Yankner BA, Shooter EM (1982) The biology and mechanism of action of nerve growth factor. Annu Rev Biochem 51: 845–868
Yasuda T, Sobue G, Mokuno K, Kreider B, Pleasure D (1987) Cultured rat Schwann cells express low affinity receptors for nerve growth factor. Brain Res 436: 113–119

The Membrane Receptor for Nerve Growth Factor

M. V. CHAO

1 Introduction

Receptors for mammalian growth factors are integral membrane proteins designed to have multiple functions for communicating with other cells and for intracellular signal transduction. These cell surface proteins have binding sites for the ligand that face the external environment and also an effector site that generates a biological signal within the cell. For nerve growth factor (NGF), the interaction with its cell surface receptor leads to a remarkable set of reactions in responsive cells, including changes in gene expression, activation of neurotransmitter enzymes, and stimulation of neuronal processes. These many events lead ultimately to the survival and maintenance of selective neuronal cell populations.

The cloning of the NGF receptor gene has provided new insights into receptor structure, expression, and distribution. As a growth factor receptor, the NGF receptor is unlike other receptors containing inherent tyrosine kinase activities. This chapter will review the structural and genetic features of the NGF receptor and its role in transmitting the intracellular signals of NGF. Several lines of evidence suggest that the signal transduction mechanism of NGF

Department of Cell Biology and Anatomy and Division of Hematology/Oncology Cornell University Medical College, 1300 York Avenue, New York, NY 10021, USA

Current Topics in Microbiology and Immunology, Vol. 165

involves interaction of the receptor with a second, auxiliary protein. The availability of genetic probes for the NGF receptor now allows investigation of the regulation of receptor expression and correlation of the structural features of the receptor with its functional consequences.

2 Cloning of the NGF Receptor Gene

Two techniques contributed tremendously to the cloning of the NGF receptor gene: (1) the use of gene transfer and (2) the ability to detect NGF receptor using monoclonal antibodies. DNA-mediated gene transfer was used as an assay for the expression cloning of the NGF receptor (CHAO et al. 1986; RADEKE et al. 1987). Stable transformation of cultured cells with total genomic DNA had previously allowed for the isolation and characterization of human oncogenes and selectable genes such as those for thymidine kinase, adenine phosphoribosyl transferase, and hypoxanthine phosphoribosyl transferase. This method circumvented the requirement for mRNA enrichment or for protein purification and thus allowed for the isolation of a gene whose product is expressed at relatively low levels.

Since expression of cell surface receptors is not amenable to metabolic selection, transfer of the gene encoding the NGF receptor required cotransfection with genes carrying selectable markers. High molecular weight DNA was mixed with the purified herpes virus thymidine kinase gene to produce a calcium phosphate precipitate. This precipitate was used to transfect mouse fibroblasts deficient in thymidine kinase (tk^-). After 2 weeks hypoxanthine-aminopterin-thymidine (HAT) selection, tk^+ colonies were screened for NGF receptor expression by an in situ rosette assay employing the ME20.4 monoclonal antibody (CHAO et al. 1986), which inhibits NGF binding to NGF receptors on human melanoma cells (ROSS et al. 1984), or by cell sorting (RADEKE et al. 1987) using the 192-IgG monoclonal antibody against the rodent receptor (CHANDLER et al. 1984). The identification and availability of these antibodies were essential to the success of the gene transfer approach.

Two different routes were taken to isolate the NGF receptor gene following gene transfer and purification of the fibroblast transfectants. The first was to use a molecular tag to rescue the phenotype-transforming sequences, and the second approach was to employ subtracted cDNA from transfected cells. Human NGF receptor sequences were isolated from a transfected mouse L cell line using human middle repetitive *Alu* sequences as a probe (CHAO et al. 1986). The rat NGF receptor gene was isolated following cDNA subtractive hybridization. Poly $(A)^+$ RNA from the L cell transfectant was converted to $[^{32}P]$cDNA, and the unique cDNA sequences were isolated after hybridization with RNA from tk^- cells. This enriched cDNA was used to identify a full-length rat NGF receptor clone (RADEKE et al. 1987).

The validity of the receptor sequence deduced from cDNA clonign was verified by gas phase sequencing of NGF receptor protein purified from the melanoma A875 cell line (MARANO et al. 1987); by cDNA expression in mammalian cells (JOHNSON et al. 1986; RADEKE et al. 1987); and by expression in *Xenopus laevis* oocytes microinjected with receptor mRNA produced in vitro (SEHGAL et al. 1988a).

The 3.8 kilobase mRNA for the NGF receptor contains a 5′ untranslated sequence of 120 bases and a large 3′ untranslated region of nearly 2 kilobases. The 3′ sequence is uninterrupted and is contained in the largest receptor exon (2.3 kilobases) together with sequences representing the intracellular domain of the receptor (see below). The significance of such an extensive 3′ untranslated region is unknown, but long noncoding sequences have been detected in the mRNAs of many genes encoding cell surface molecules. The receptor mRNA contains a single polyadenylation sequence, ATTAAA, nine bases upstream from a poly(A) tail. No other polyadenylation sites have been uncovered for the receptor mRNA.

3 Structural Features of the NGF Receptor

Complementary cDNAs representing the NGF receptor mRNA have been isolated for chicken, rat, and humans by screening cDNA libraries from chicken brain (HEUER et al. 1990; LARGE et al. 1989), transfected fibroblasts (RADEKE et al. 1987), and human A875 melanoma cells (JOHNSON et al. 1986). Full-length receptor cDNAs predict a signal sequence of 28 amino acids; an extracellular domain containing four 40 amino acid repeats with six cysteine residues at conserved positions, followed by a region rich in threonine and serine residues; a single transmembrane domain; and a 155 amino acid cytoplasmic domain. The molecular weight of the fully processed receptor protein is approximately 70–80 000. The difference between the size of the mature receptor and the predicted size of the receptor protein (399 amino acids) is due to extensive O- and N-linked glycosylation (see Ross, this volume).

A comparison of the amino acid sequences of the human, rat, and chicken NGF receptors is displayed in Fig. 1. The rat NGF receptor sequence (RADEKE et al. 1987) shares 92% homology on the amino acid level with the human receptor, whereas the chicken NGF receptor is less conserved (HEUER et al. 1990; LARGE et al. 1989). It is noteworthy that nearly every cysteine residue is conserved and that the transmembrane region is especially well-conserved between the three species.

Most of the differences in amino acid sequence between the human and rat receptors are conservative changes. However, several differences which change the net charge of the receptor may account for the extreme species specificity of monoclonal antibodies against the receptor. For example, some of the monoclonal antibodies against the human receptor are effective at blocking the binding

```
                                   +1                  20                  40
                    Human          KEACPTGLYTHSGECCKACNLGEGVAQPCGANQTVCEPCL
                    Rat            --T-S-----------------------------------
                    Chicken                  T----Y---------V----V---------

                   60                  80                 100
DSVTFSDVVSATEPCKPCTECVGLQSMSAPCVEADDAVCRCAYGYYQDETTGRCEACRVCEAGSGLVFSCQ
-N-------------------L---------------------------E--H----S---V---------
----Y--T-----------Q----H--------S-----------F---LS-T-KE-TI--V-F--M-P-R

        120                 140                 160                 180
DKQNTVCEECPDGTYSDEANHVDPCLPCTVCEDTERQLRECTRWADAECEEIPGRWITRSTPPEGSDSTAP
-----------E------------------------------P--------------P-------------
-S-D-------E-SF-----F--------I--EN-VMVK--KATS----RDLHP--T-HTPSLA----PE-

                 200                 220                 240
STQEPEAPPEQDLIASTVAGVVTTVMGSSQPVVTRGTTDNLIPVYCSILAAVVVGLVAYIAFKRWNSCKQN
------V------VP----DM--------------------------------------------------
I-RD-FNT---EGM-T-L-DI------------S---A---------------------------------

       260                 280                 300                 320
KQGANSRPVNQTPPPEGEKLHSDSGISVDSQSLHDQQPHTQTASGQALKGDGGLYSSLPPAKREEVEKLLN
-------------------------------------T--------------N------LT----------
-----N-------S------------------------PN-STQ-P-P----S--A----S-Q-------S

               340                 360                 380
GSAGDTWRHLAGELGYQPEHIDSFTHEACPVRALLASWATQDSATLDALLAALRRIQRADLVESLCSESTATSPV
-xxx----------------------------------GA--------------------I--------------
SSAEE---Q-------KEDL--C--R-ES-A-----D-SAKET-------V---K---G-IA---Y-------xV
```

Fig. 1. Comparison of the human and rat NGF receptor sequences. Identical residues between the chicken (HEUER et al. 1990), rat (RADEKE et al. 1987), and human (JOHNSON et al. 1986) NGF receptor sequence are shown with a *dash*. The *boxed region* represents the putative transmembrane domain. The NH_2-terminal amino acid lysine is designated as the first amino acid

```
                     M G A G A T G R A M D G P R L L L L L L L G V S L G G A
K E A C P T G L Y           T H S G E C   C K A C N L G E G V A Q P C G A N   Q T V C E     I
    P C L D S V T F S D L V S A T E P C K P C T E C V G L Q S M S A P C V E A D D A V C R     II
      C A Y G Y Y Q D E         T T G R C E A C R V C E A G S G L V F S C Q D K Q N T V C E E   III
      C P D G T Y S D E     A N H V D P C L P C T V C E D T E R Q L R E C T R W A D A E C E E   IV
```

Fig. 2. The NGF receptor contains four internal repeats. The signal peptide and the first 160 amino acids of the human NGF receptor are displayed. This region has been divided into four segments of approximately 40 amino acids. Spacing has been introduced to allow for maximal alignment of the 24 cysteine residues in this domain

of NGF to the human receptor; however, they do not recognize the rat or chicken receptor. Undoubtedly, the differences in amino acid sequence in the putative extracellular domain contribute to the species specificity of the anti-rat and -human monoclonal antibodies.

One of the most striking features of the receptor amino acid sequence are the 28 cysteine residues, 24 of which are found within the first 160 residues of the receptor (Fig. 2). Alignment of the cysteine residues indicates that they are contained in four 40 amino acid internal repeats that have many similar residues besides the cysteine residues. These internal repeats have lead to the hypothesis that these units are evolutionarily related. Similar cysteine rich repeats have been detected in the extracellular domains of a number of proteins including cell surface proteins such as the tumor necrosis factor receptor (LOETSCHER et al. 1990; SCHALL et al. 1990; SMITH et al. 1990), CDw40 (STAMENKOVIC et al. 1989), T cell antigens (KWON and WEISSMAN 1989; MALLET et al. 1990) and a Shope viral protein (UPTON et al. 1987).

The pattern of cysteine repeats found in these cell membrane proteins is shown in Fig. 3. It is likely that the NGF receptor is oriented with the NH_2-terminal region outside the cell, and the COOH-terminal in the cytoplasm. Since the cysteine repeats constitute 70% of the extracellular portion of the NGF receptor, this domain probably constitutes part or all of the binding site for NGF.

The extracellular sequence is also distinguished by its highly negative charge. The bulk of the negative charge is contributed by the cysteine rich domain in which the net charge is −24, consistent with the acidic pI of the receptor (GROB et al. 1985). No charged amino acids are found in residues 223–244 which have all the characteristics of a transmembrane domain. This hydrophobic segment is followed by several basic amino acids which commonly mark the cytoplasmic border of the membrane spanning region. Receptors for asialoglycoprotein, transferrin, EGF, interleukin-1, interleukin-2, interleukin-6, and platelet derived growth factor (PDGF) also contain a single transmembrane domain.

```
NGFR   G E C C K A C N L G E G V A Q P C G A N Q   T V C E P C L D N V T
SFVT2  G L C C A S C H P G F Y A S R L C G P G S N T V C S P C E D G T F
TNFR1  Q M C C S K C S P G Q H A K V F C T K T S D T V C D S C E D S T Y
TNFR2  S I C C T K C H K G T Y L Y N D C P G P G Q T D C R E C   E S G S
CDw40  S Q C C S L C Q P G Q K L V S D C T E F T E T E C L P C   G E S E
OX40   H K C C R E C Q P G H G M V S R C D H T R D T V C H P C   E P G F
```

Fig. 3. Internal cysteine rich repeats for cell surface proteins. Displayed are the arrangement of cysteines in one of the repeats from the NGF receptor (NGFR), receptors for TNF (TNFR), Shope fibroma virus T2 gene (SFVT2), B cell antigen CDw40, and T cell antigen OX40

Though the deduced amino acid sequence of the human NGF receptor reveals structural features in common with other receptors whose structures have been elucidated, the NGF receptor lacks significant amino acid similarities with any other known receptor or protein. In particular, the primary sequence of the intracellular domain does not display any similarities to known growth factors, oncogenes, or tyrosine kinase gene family. This lack of homology to known tyrosine kinases is consistent with the absence of tyrosine kinase activity by the NGF receptor from human melanoma cells (GROB et al. 1985), PC12 cells, and rat superior cervical ganglia (TANIUCHI et al. 1986a) and in spite of reports that tyrosine phosphorylation occurs in NGF-treated PC12 cells (MAHER 1988). The lack of similarity between the NGF receptor and the tyrosine kinase family of receptors is not unexpected since the primary target cells influenced by NGF do not undergo cell division in response to the hormone. The absence of such activity for NGF receptor suggests that the mechanism of action of NGF differs significantly from other peptide growth factors such as PDGF, EGF, colony stimulating factor 1 (CSF-1), and insulin which are distinguished by the inherent tyrosine kinase activities of their receptors. Hence, the role of the 155 amino acid cytoplasmic domain in NGF receptor function is not yet known, but this domain does not contain any enzymatic activities such as serine or tyrosine kinase activities.

4 Receptor Gene Structure

Somatic cell hybridization established that the human NGF receptor gene resides on the long arm of chromosome 17 (HUEBNER et al. 1986). The monoclonal antibody against the human receptor was used independently to assign the receptor gene to chromosome 17 (RETTIG et al. 1986). Using genomic probes for the receptor gene, BREAKEFIELD et al. (1986) have demonstrated an absence of linkage to familial dysautonomia, a disease with defects that suggest NGF deprivation. Similarly, the NGF receptor gene locus was also used to establish that the genetic locus accounting for von Recklinghausen's neurofibromatosis was near the centromeric region of chromosome 17 (SEIZINGER et al. 1987). Furthermore, it was shown that the NGF receptor was not the causal gene defect.

The genomic organization of the receptor gene was established by analysing overlapping bacteriophage clones containing the receptor sequence (SEHGAL et al. 1988b). The structural features of the NGF receptor were found to be reflected in the organization of the gene (Fig. 4). Analysis of the splice junction sites for the six exons of the receptor gene revealed that structurally distinct domains of the receptor are in separate exons (SEHGAL et al. 1988b). For example, the second and third exons encode the NH_2-terminal amino acid sequence of the mature receptor protein and a discrete domain of 161 amino acids, of which 24 are cysteine residues (Table 1). It is of interest that the splice site for the third exon is located precisely after the 24th cysteine residue. Similarly, the transmembrane

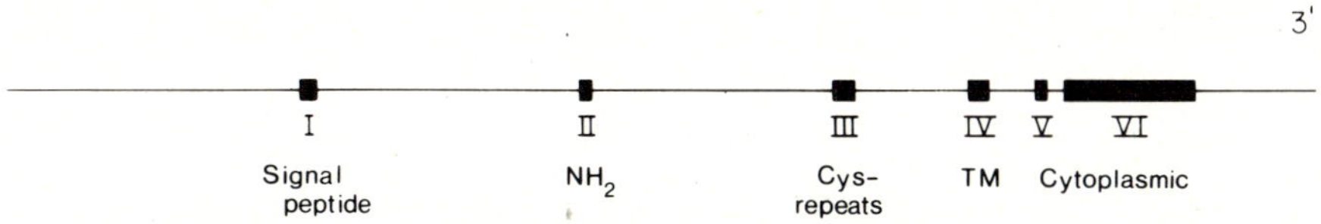

Fig. 4. Genomic structure of the human NGF receptor gene. The NGF receptor gene is shown in a 5′ to 3′ orientation with exons represented as *boxes*. The intron-exon organization of the NGF receptor gene was defined by DNA sequence analysis (SEHGAL et al. 1988b)

region found in the fourth exon contains an intron-exon boundary in the stop transfer sequence directly following the hydrophobic containing amino acids. Such precise divisions of the structural features of the receptor molecule are supportive of a functional role of these domains.

The initiation site for NGF receptor gene transcription in human A875 melanoma cells and in rat PC12 cells was localized to 122 nucleotides upstream of the initiating methonine residue. Both primer extension and S1 nuclease protection experiments were used to verify the initiation site (SEHGAL et al. 1988b). The upstream promoter sequence was found to be capable of high levels of expression when placed upstream of an unrelated reporter gene, such as chloramphenicol acetyltransferase.

The sequence of the receptor promoter reveals a high G + C content but lacks consensus TATA and CAAT sequences. Several conserved GGGCGG sequences are present within 20 nucleotides of the initiator ATG codon (Fig. 5). These GC rich sequences represent potential binding sites for the transcription factor Sp1 (DYNAN and TJIAN 1983). Several eukaryotic genes have promoters which are rich in GC content but lack the characteristic TATA and CAAT boxes. Most of these genes encode enzymes with housekeeping functions and expression of these genes is at a low constitutive level in diverse tissues. Significantly, the promoters of several cellular growth control genes such as c-Harvey *ras* (ISHII et al. 1985a), c-Kirsten *ras* (HOFFMAN et al. 1987), the EGF receptor (ISHII et al. 1985b), and the insulin receptor (ARAKI et al. 1987) also lack the typical TATA and CAAT transcriptional consensus sequence, but have multiple GC box motifs. Since these gene products are in some way involved with growth control and their promoter elements are similar, regulation of these genes may share common mechanisms.

Table 1. Exons of the human NGF receptor gene encode structurally significant domains

Exon	Domain	Size (base pairs)
I	Signal peptide	179
II	NH_2-terminal	141
III	Cysteine rich	360
IV	Transmembrane	255
V	Cytoplasmic	159
VI	COOH-terminal 3′ untranslated	2310

GCAGCCGGAGAGAGGAACAGGAACCGCAGTGGACGCCCCTGGTCCCCGGGACGACGCGCAGTAGCGC
-250 -230 -210

CGGGAACTGGGTACCA**GGGCGGG**TGGGTGAGAGGCTCTAAGGGACAAGGCAGGGAGAAGCGCACGGGT
-190 -170 -150

GCGGGGAACCAGCCCTCCCTTTGCCTCTGCTTCCCACCCCGAGGCGGCA**GGGCGGG**CGGGCAGGTTC
-130 -110 -90 -70

CGGGGGT**GGCGGG**CTGGCG**GGGCGG**AGGCGGGGGCCGACAGCTGGCTTCACCAGCCTTCCGCCGTAG
-50 -30 -10

Fig. 5. The NGF receptor promoter sequence. The initiation site for NGF receptor transcription was mapped by primer extension and S1 nuclease protection analysis (SEHGAL et al. 1988b). *Boldface* nucleotides denote potential sites of interaction with the transcription factor Sp1. The numbering begins from the site of initiation (+1)

The promoter for the NGF receptor therefore resembles those genes which are constitutively expressed or involved in growth regulation. An unusual features of this promoter is that it appears to direct transcription primarily from a major site of initiation instead of using multiple sites of initiation as occurs with many constitutively expressed genes. The single initiation site was observed in both human melanoma cells and PC12 cells (SEHGAL et al. 1988b). The constitutive nature of the receptor promoter implies that the neuronal-specific expression and developmental regulation of the receptor gene is dependent upon other regulatory elements. Transgenic mice expressing human NGF receptors in appropriate cell types indicate that neuronal specific expression is dictated by DNA sequences apart from the promoter (PATIL et al. 1990). In addition, the features of the receptor promoter may explain not only why this tissue-specific gene is expressed efficiently in mouse fibroblasts after transfection with genomic DNA, but may also account for the detection of NGF receptors in nonneuronal cells (RAIVICH et al. 1985; THORPE et al. 1987).

5 Expression of NGF Receptors

Transcription of the NGF receptor gene is undoubtedly necessary for the continued maintenance and survival of sympathetic and sensory nerve systems as well as cholinergic neurons in the basal forebrain (GNAHN 1986). As described elsewhere in this volume, NGF receptors have been primarily characterized by cross-linking with radioiodinated NGF at the cell surface (MASSAGUE et al. 1981; GROB et al. 1983; HOSANG and SHOOTER 1985; GREEN and GREENE 1986); immunoreactivity with monoclonal antibodies (ROSS et al. 1984; SPRINGER et al. 1986; GARIN CHESA et al. 1988); equilibrium binding to whole cells (SUTTER et al. 1979; SCHECHTER and BOTHWELL 1981); and radioautography of iodinated NGF bound to tissue sections (RICHARDSON et al. 1986; BERND 1986; RAIVICH and

KREUTZBERG 1987). The availability of recombinant DNA clones for the chicken, rat, and human NGF receptors provides additional tools for investigating the regulation of NGF receptor gene expression during development of the nervous system. In particular, in situ hybridization (ERNFORS et al. 1988; AYER-LELIEVRE et al. 1988a; GIBBS et al. 1989), RNA blot analysis, and S1 nuclease protection (BUCK and CHAO 1989) have been useful techniques.

Classic target cell populations for NGF have been examined for in vivo expression of NGF receptor mRNA (BUCK et al. 1987; SCHATTEMAN et al. 1988; ERNFORS et al. 1988). In rat sympathetic ganglia, which utilize NGF throughout the lifespan of the animal steady state levels of NGF receptor mRNA increase with development. In contrast, dorsal root ganglia sensory neurons that require NGF for survival only during early developmental periods express lower levels of receptor mRNA levels after birth (BUCK et al. 1987). Receptor mRNA levels reach a peak in early chicken brain development at embryonic days 6–10, but decrease significantly at later embryonic ages (ESCANDON and CHAO 1989). These observations indicate that the appearance and disappearance of the NGF receptor in these neuronal tissues may be a significant parameter that determines the ability of NGF to influence selective cell populations. Since NGF appears to be synthesized at a low, constant level in most tissues, regulation of receptor affinity and gene expression may determine which cells respond to NGF.

Perhaps the most interesting aspect of receptor expression has been nonneuronal cell populations, such as testis (AYER-LELIEVRE et al. 1988b), and Schwann cells after peripheral nerve transection (TANIUCHI et al. 1986b; HEUMANN et al. 1987) or cell culture (LEMKE and CHAO 1988). The increased levels of mRNA and protein for both NGF and NGF receptor in Schwann cells after nerve lesioning are not understood, but this event may facilitate nerve regeneration (JOHNSON et al. 1988). The expression of NGF receptor mRNA in areas such as testis, muscle (RAIVICH et al. 1985), and lymphoid tissues such as thymus and spleen (ERNFORS et al. 1988) indicates that the actions of NGF may not be restricted only to cells derived from the neural crest. For example, NGF is known to bind to T cells (THORPE et al. 1987), to influence the differentiation of eosinophils and basophils (MATSUDA et al. 1988), and to have proliferative effects upon mast cells (ALOE and LEVI-MONTALCINI 1977).

6 Transfection of NGF Receptors

Equilibrium binding of ^{125}I-NGF to responsive cells reveals two distinct affinity states for the NGF receptor (SUTTER et al. 1979; LANDRETH and SHOOTER 1980; SCHECHTER and BOTHWELL 1981). In most responsive cells, such as neurons and PC12 cells (GREENE and TISCHLER 1976), approximately 10%–15% of the receptors display high affinity binding with a K_d of 10^{-11} *M*, with the remainder of the receptors possessing a K_d of 10^{-9} *M*. The difference in equilibrium binding is accounted for by a 100-fold difference in the rate of dissociation of NGF. It is generally believed that the high affinity receptor is internalized upon NGF binding and is responsible for the actions of NGF.

Human and rat NGF receptor cDNAs introduced into mouse fibroblast cell lines only give rise to the low affinity kinetic class of receptors (JOHNSON et al. 1986; RADEKE et al. 1987). As many as 500 000 receptors per cell can be observed after transfection; however, none of the cell lines display any detectable response to NGF treatment (HEMPSTEAD et al. 1988).

The introduction of a full length human receptor cDNA into a variant PC12 cell line (NR18) resulted in the generation of both high and low affinity receptors with affinities similar to those of the kinetic classes found in sympathetic neurons and PC12 cells (HEMPSTEAD et al. 1989). The NR18 cell line was chosen due to the lack of endogenous NGF receptors and the inability to respond to NGF (BOTHWELL et al. 1980). The appearance of high affinity human NGF receptors in this cell line was correlated with a functional response, the induction of c-*fos* transcription by NGF. These results indicate that the cloned receptor cDNA can give rise to functional NGF receptors in appropriate cells. The detection of only one mRNA species in cell lines that display only low affinity receptors (melanoma and transfected fibroblasts) and cell lines that have predominantly high affinity receptors (SY-5Y neuroblastoma) also demonstrates that one gene give rise to a single mRNA species, specifying both kinetic forms of the NGF receptor.

The lack of appropriate responses to NGF in many cells expressing abundant NGF receptors strongly suggests that most cells do not possess either the correct cellular environment or the appropriate signaling machinery. The lack of a response can also be correlated with the absence of the high affinity form of the receptor, which mediates many of the biological responses of NGF (GREEN et al. 1986). Hence, clarification of the biochemical nature of the high affinity form is crucial in understanding the mechanisms by which the receptor protein transmits its intracellular signal after binding to NGF. Differential splicing does not appear to give rise to the two kinetic forms of the receptor and, furthermore, no other evidence has been found for another gene related to the receptor that binds NGF (HEMPSTEAD et al. 1988). Although posttranslational modifications may account for a difference of affinity, there is presently no indication that a posttranslational even influences ligand affinity for NGF.

The most plausible explanation for the difference in receptor ligand affinity for the NGF receptor is that a distinct regulatory protein interacts with the receptor and dictates the affinity of the receptor for its ligand. The expression of high affinity receptors only in selected cells from the neural crest implies that a separate protein may be required for the high affinity state. Several different observations makes this a tenable hypothesis. First, biochemical attempts to characterize the NGF receptor using affinity cross-linking agents have described cross-linked species larger than the 100 000 molecular weight receptor-NGF complex. A higher molecular weight species of 135 000–158 000 can be identified by cross-linking ^{125}I-NGF with the receptor using a lipophilic reagent (MASSAGUE et al. 1981; HOSANG and SHOOTER 1985; GREEN and GREENE 1986).

This complex is though to represent ^{125}I-NGF cross-linked with the NGF receptor together with a separate regulatory protein.

Second, examination of the human (JOHNSON et al. 1986), rat (RADEKE et al. 1987), and chicken (LARGE et al. 1989; HEUER et al. 1990) NGF receptor amino acid sequences indicates that an extremely strong region of conservation is present in the transmembrane region and extends 50 amino acids into the cytoplasmic domain (Fig. 1). No differences exist in any of the amino acids of the chicken, rat, and human NGF receptors in this transmembrane region. It is attractive to speculate that this particular domain of the receptor is so highly conserved because of important interactions with other proteins closely associated with the plasma membrane. There are a number of amino acid changes throughout the molecule, and these differences do not appear to inhibit the function of the receptor. For example, the binding domain for NGF must be highly conserved in the receptor structure, since the mouse submaxillary NGF that has been used almost exclusively for equilibrium binding studies gives the same values of K_d for receptors on human, rat, bovine, or chicken cells. Additionally, receptors from different species for interleukin-2 (SHIMUZU et al. 1985) and LDL (YAMAMOTO et al. 1986) contain differences in amino acid sequence in transmembrane domains, suggesting that the lack of change in this transmembrane domain for the NGF receptor must be highly significant and essential for the function of the receptor.

Thus, the two kinetic forms of the NGF receptor appear to be encoded by the same protein (GREEN and GREENE 1986) which is the product of a single gene (HEMPSTEAD et al. 1989). Several possibilities remain for an associated protein which is capable of modulating the affinity state of the NGF receptor. Microinjection or introduction of *ras* (BAR-SAGI and FERAMISCO 1985; HAGAG et al. 1986; NODA et al. 1985; GUERRERO et al. 1986) or *sarc* (ALEMA et al. 1986) oncogene products leads to a differentiation of PC12 cells or sympathetic neurons (BOROSCIA et al. 1989) much like the effect of NGF. It remains to be established whether NGF-mediated events are directly dependent upon interactions of the receptor with *ras* or *sarc* proteins. The structure of the receptor protein does not contain any sequences which interact with known G proteins. Resolution of the molecular basis of the two kinetic forms of the NGF receptor will undoubtedly begin to unravel the complex series of events that result from NGF binding to responsive cells.

7 Conclusions

The unique set of responses elicited by NGF and its role as a trophic and tropic factor implies that the signal transduction mechanism is distinctive. From results of transfection of cDNA clones for the receptor and biochemical experiments using affinity cross-linking with ^{125}I-NGF, the high affinity functional form of the NGF receptor most likely represents a bimolecular complex consisting of the

80 000–85 000 molecular weight binding protein in association with an accessory protein. The ability to generate a responsive receptor in certain populations of cells is reminiscent of the behavior of the interleukin-2 (IL-2) receptor, which displays high affinity IL-2 binding in appropriate T cell populations. The IL-2 receptor contains a second accessory molecule that binds IL-2. Unlike this lymphokine receptor, however, the presumed accessory molecule associated with the NGF receptor does not bind NGF (Hempstead et al. 1989).

Despite many attempts to identify a second messenger for NGF, the precise role and identity of signaling pathways for NGF have not been established. The accumulated data indicate that NGF binding to its receptor does not utilize cAMP as a second messenger, and does not elicit a change in either Ca^{+2} mobilization or turnover of phosphoinositides (Chao 1990). On one level, it is clear that protein phosphorylation events mediate the many effects that NGF has upon responsive cells (see Halegoua et al., this volume). Dissection of the steps in the signal transduction pathways of NGF will undoubtedly require the molecular and biochemical characterization of the high affinity NGF receptor and an understanding of its role in carrying signals from the cell membrane to the interior of the cell.

References

Alema S, Casalbore P, Agostini E, Tato F (1985) Differentiation of PC12 phaeochromocytoma cells induced by v-*src* oncogene. Nature 316: 557–559

Aloe L, Levi-Montalcini R (1977) Mast cells increase in tissues of neonatal rats injected with nerve growth factor. Brain Res 133: 356–366

Araki E, Shimada F, Uzawa H, Mori M, Ebina Y (1987) Characterization of the promoter region of the human insulin receptor gene. J Biol Chem 262: 16186–16191

Ayer-LeLievre C, Olson L, Ebendal T, Seiger A, Persson H (1988a) Expression of the β-nerve growth factor gene in hippocampal neurons. Science 240: 1339–1341

Ayer-LeLievre C, Olson L, Ebendal T, Hallbook F, Persson H (1988b) Nerve growth factor mRNA and protein in the testis and epididymis of mouse and rat. Proc Natl Acad Sci USA 85: 2628–2632

Bar-Sagi D, Feramisco JR (1985) Microinjection of the ras oncogene protein into PC12 cells induces morphological differentiation. Cell 42: 841–848

Bernd P (1985) Appearance of nerve growth factor receptors on cultured neural crest cells. Dev Biol 112: 145–156

Bothwell M, Schechter AL, Vaughn KM (1980) Clonal variants of PC12 pheochromocytoma cells with altered response to nerve growth factor. Cell 21: 857–866

Borasio GD, John J, Wittinghofer A, Barde Y-A, Sendtner M, Heumann R (1989) Ras p21 protein promotes survival and fiber outgrowth of cultured embryonic neurons. Neuron 2: 1087–1096

Breakefield XO, Ozelius L, Bothwell MA, Chao MV, Axelrod F, Kramer PL, Kidd KK, Lanahan AA, Johnson DE, Ross AH, Gusella JF (1986) DNA polymorphism for the nerve growth factor receptor gene excludes its role in familial dysautonomia. Mol Biol Med 3: 483–494

Buck CR, Chao MV (1989) The NGF receptor: molecular approaches. In: Rush R (ed) IBRO handbook of the neurosciences. Wiley, New York, pp 327–348

Buck CR, Martinez H, Black IB, Chao MV (1987) Developmentally regulated expression of the nerve growth factor receptor gene in the periphery and brain. Proc Natl Acad Sci USA 84: 3060–3063

Chandler CE, Parsons LM, Hosang M, Shooter EM (1984) A monoclonal antibody modulates the interaction of nerve growth fctor with PC12 cells. J Biol Chem 259: 6882–6889

Chao MV, Bothwell MA, Ross AH, Koprowski H, Lanahan A, Buck CR, Sehgal A (1986) Gene transfer and molecular cloning of the human NGF receptor. Science 232: 418–421

Chao MV (1990) Nerve growth factor. In Peptide Growth Factors and their Receptors. Springer-Verlag: Berlin, pp 135–166

Dynan WS, Tjian R (1983) The promoter-specific transcription factor Sp1 binds to upstream sequences in the SV40 early promoter. Cell 35: 79–87

Ernfors P, Hallbook F, Ebendal T, Shooter EM, Radeke MJ, Misko TP, Persson H (1988) Developmental and regional expression of β-nerve growth factor receptor messenger RNA in the chick and rat. Neuron 1: 983–996

Escandon E, Chao MV (1989) Developmental expression of the chicken nerve growth factor receptor gene during brain morphogenesis. Dev Brain Res 47: 187–196

Garin Chesa P, Rettig WJ, Thomson TM, Old LJ, Melamed MR (1988) Immunohistochemical analysis of nerve growth factor expression in normal and malignant human tissues. J Histochem Cytochem 36: 383–389

Gibbs RB, McCabe JT, Buck CR, Chao MV, Pfaff DW (1989) Expression of NGF receptor in the rat forebrain detected with in situ hybridization and immunohistochemistry. Mol Brain Res 6: 257–287

Gnahn H, Hefti F, Heumann R, Schwab ME, Thoenen H (1983) NGF-mediated increase of choline acetyltransferase (ChAT) in the neonatal rat forebrain: Evidence for a physiological role of NGF in the brain? Dev Brain Res 9: 45–52

Green SH, Greene LA (1986) A single $M_r = 103{,}000$ ^{125}I-β-nerve growth factor-affinity-labeled species represents both the low and high affinity forms of the nerve growth factor receptor. J Biol Chem 261: 15316–15326

Greene LA, Tischler (1976) Establishment of a noradrenergic clonal line of rat adrenal pheochromocytoma cells which respond to nerve growth factor. Proc Natl Acad Sci USA 73: 2424–2428

Green SH, Rydel RE, Connolly JL, Greene LA (1986) PC12 cell mutants that possess low- but not high-affinity nerve growth factor receptors neither respond to nor internalize nerve growth factor. J Cell Biol 102: 830–843

Grob PM, Berlot CH, Bothwell MA (1983) Affinity labeling and partial purification of nerve growth factor receptors from rat pheochromocytoma and human melanoma cells. Proc Natl Acad Sci USA 80: 6819–6823

Grob PM, Ross AH, Koprowski H, Bothwell MA (1985) Characterization of the human melanoma nerve growth factor receptor. J Biol Chem 260: 8044–8049

Guerrero I, Wong H, Pellicer A, Burstein DE (1986) Activated N-*ras* gene induces neuronal differentiation of PC12 rat pheochromocytoma cells. J Cell Physiol 129: 71–76

Hagag N, Halegoua S, Viola M (1986) Inhibition of growth factor-induced differentiation of PC12 cells by microinjection of antibody to ras p21. Nature 319: 680–682

Hempstead BL, Schleifer LS, Chao MV (1989) Expression of functional nerve growth factor receptors after gene transfer. Science 243: 373–375

Hempstead BL, Patil N, Olson K, Chao M (1989) Molecular analysis of the nerve growth factor receptor. Cold Spring Harbor Symp Quant Biol 53: 477–485

Heuer JG, Fatemie-Namie S, Wheeler EF, Bothwell M (1990) Structure and developmental expression of the chicken NGF receptor. Dev Biol 137: 287–304

Heumann R, Lindholm D, Bandtlow C, Meyer M, Radeke MJ, Misko TP, Shooter E, Thoenen H (1987) Differential regulation of mRNA encoding nerve growth factor and its receptor in rat sciatic nerve during development, degeneration, and regeneration: role of macrophages. Proc Natl Acad Sci USA 84: 8735–8739

Hoffman EK, Trusko SP, Freeman N, George DL (1987) Structural and functional characterization of the promoter region of the mouse c-Ki-*ras* gene. Mol Cell Biol 7: 2592–2596

Hosang M, Shooter EM (1985) Molecular characteristics of nerve growth factor receptors on PC12 cells. J Biol Chem 260: 655–662

Huebner K, Isobe M, Chao M, Bothwell M, Ross AH, Finan J, Hoxie JA, Sehgal A, Buck CR, Lanahan A, Nowell PC, Koprowski H, Croce CM (1986) The nerve growth factor receptor gene is at human chromosome region 17q12-17q22, distal to the chromosome 17 breakpoint in acute leukemias. Proc Natl Acad Sci USA 83: 1403–1407

Ishii S, Merlino GT, Pastan I (1985a) Promoter region of the human Harvey *ras* proto-oncogene: similarity to the EGF receptor proto-oncogene promoter. Science 230: 1378–1381

Ishii S, Xu YH, Stratton BA, Roe GT, Merlino GT, Pastan I (1985b) Characterization and sequence of the promoter region of the human epidermal growth factor receptor gene. Proc Natl Acad Sci USA 82: 4902–4904

Johnson D, Lanahan T, Buck CR, Sehgal A, Morgan C, Mercer E, Bothwell M, Chao M (1986) Expression and structure of the human NGF receptor: Cell 47: 445–454

Johnson EM, Taniuchi M, DiStefano PS (1988) Expression and possible function of nerve growth factor receptors on Schwann cells. Trends Neurosci 11: 299–304

Kwon BS, Weissman SM (1989) cDNA sequences of two inducible T-cell genes. Proc Natl Acad Sci USA 86: 1963–1967

Landreth GE, Shooter EM (1980) Nerve growth factor receptors on PC12 cells: ligand-induced conversion from low- to high-affinity states. Proc Natl Acad Sci USA 77: 4751–4755

Large TH, Weskamp G, Helder JC, Radeke MJ, Misko TP, Shooter EM, Reichardt LF (1989) Structure and developmental expression of the nerve growth factor receptor in the chicken central nervous system. Neuron 2: 1123–1134

Lemke G, Chao M (1988) Axons regulate Schwann cell expression of major myelin and NGF receptor genes. Development 102: 499–504

Loetscher H, Pan Y-CE, Lahm H-W, Gentz R, Brockhaus M, Tabuchi H, Lesslauer W (1990) Molecular cloning and expression of the human 55 kd tumor necrosis factor receptor. Cell 61: 352–359

Maher PA (1988) Nerve growth factor induces protein-tyrosine phosphorylation. Proc Natl Acad Sci USA 85: 6788–6791

Mallet S, Fossum S, Barclay AN (1990) Characterization of the MRC OX40 antigen of antivated CD4 T lymphocytes—a molecule related to nerve growth factor receptor. EMBO 9: 1063–1068

Marano N, Dietzschold B, Earley JJ, Schatteman G, Grob P, Ross AH, Bothwell M, Koprowski H (1987) Purification and amino terminal sequencing of human melanoma nerve growth factor receptor. J Neurochem 48: 225–232

Massague J, Guillette BJ, Czech MP, Morgan CJ, Bradshaw RA (1981) Identification of a nerve growth factor receptor protein in sympathetic ganglia membranes by affinity labeling. J Biol Chem 256: 9419–9424

Matsuda H, Coughlin MD, Bienenstock J, Denburg JA (1988) Nerve growth factor promotes human hemopoietic colony growth and differentiation. Proc Natl Acad Sci USA 85: 6508–6512

Noda M, Ko M, Ogura A, Liu DG, Amano T, Takano T, Ikawa Y (1985) Sarcoma viruses carrying *ras* oncogenes induce differentiation-associated properties in a neuronal cell line. Nature 318: 73–75

Patil N, Lacy L, Chao MV (1990) Specific neuronal expression of human NGF receptors in the basal forebrain and cerebellum of transgenic mice. Neuron 2: 437–447

Radeke MJ, Misko TP, Hsu C, Herzenberg LA, Shooter EM (1987) Gene transfer and molecular cloning of the rat nerve growth factor receptor. Nature 325: 593–597

Raivich G, Kreutzberg GW (1987) The localization and distribution of high affinity β-nerve growth factor binding sites in the central nervous system of the adult rat. A light microscopic autoradiographic study using ^{125}I-nerve growth factor. Neuroscience 20: 23–36

Raivich G, Zimmermann A, Sutter A (1985) The spatial and temporal pattern of βNGF receptor expression in the developing chick embryo. EMBO J 4: 637–644

Rettig WJ, Thomson TM, Spengler BA, Biedler JL, Old LJ (1986) Assignment of human nerve growth factor receptor gene to chromosome 17 and regulation of receptor expression in somatic cell hybrids. Somatic Cell Mol Genet 12: 441–447

Richardson PM, Verge Issa VMK, Riopelle RJ (1986) Distribution of neuronal receptors for nerve growth factor in the rat. J Neurosci 6: 2312–2321

Ross AH, Grob P, Bothwell MA, Elder DE, Ernst CS, Marano N, Ghrist BFD, Slemp CC, Herlyn M, Atkinson B, Koprowki H (1984) Characterization of nerve growth factor receptor in neural crest tumors using monoclonal antibodies. Proc Natl Acad Sci USA 81: 6681–6685

Schall TJ, Lewis M, Koller KJ, Lee A, Rice GC, Wong GHW, Gatanaga T, Granger GA, Lentz R, Raab H, Kohr WJ, Goddel V (1990) Molecular cloning and expression of a receptor for human tumor necrosis factor. Cell 61: 361–370

Schatteman GC, Gibbs L, Lanahan AA, Claude P, Bothwell M (1988) Expression of NGF receptor in the developing and adult primate central nervous system. J Neurosci 8: 860–873

Schechter AL, Bothwell MA (1981) Nerve growth factor receptors on PC12 cells: evidence for two receptor classes with differing cytoskeletal association. Cell 24: 867–874

Sehgal A, Wall D, Chao M (1988a) Efficient processing and expression of human nerve growth factor receptors in *Xenopus laevis* oocytes: effects on maturation. Mol Cell Biol 8: 2242–2246

Sehgal A, Patil N, Chao MV (1988b) A constitutive promoter directs expression of the nerve growth factor receptor gene. Mol Cell Biol 8: 3160–3167

Seizinger BR, Rouleau GA, Ozelius LJ, Lane AH, Faryniarz AG, Chao MV, Huson S, Korf BR, Perry DM, Pericak-Vance MA, Collins FS, Hobbs WJ, Falcone BG, Iannazzi JA, Roy JC, St George-Hyslop PH, Tanzi RE, Bothwell MA, Upadhyaya M, Harper P, Goldstein AE, Hoover DL, Bader JL, Spence MA, Mulvihill JJ, Aylsworth AS, Vance JM, Rossenwasser GOD, Gaskell PC, Roses AD, Martuza RL, Breakefield XO, Gusella JF (1987) Genetic linkage of von Recklinghausen neurofibromatosis to the nerve growth factor receptor gene. Cell 49: 589–594

Shimuzu A, Kondo S, Takeda S, Yodoi J, Ishida N, Sabe H, Osawa S, Diamantstein T, Nikaido T, Honjo T (1985) Nucleotide sequence of mouse IL-2 receptor cDNA and its comparison with the human IL-2 receptor sequence. Nucleic Acids Res 13: 1505–1516

Smith CA, Davis T, Anderson D, Solam L, Beckmann MP, Jerzy R, Dower SK, Cosman D, Goodwin RG (1990) A receptor for tumor necrosis factor defines an unusual family of cellular and viral proteins. Science 248: 1019–1023

Springer JE, Koh S, Tayrien MW, Loy R (1986) Basal forebrain magnocellular neurons stain for nerve growth factor receptor: correlation with cholinergic cell bodies and effects of axotomy. J Neurosci Res 17: 111–118

Stamenkovic I, Clark EA, Seed B (1989) A B-lymphocyte activation molecule related to the nerve growth factor receptor and induced by cytokines in carcinomas. EMBO J 8: 1403–1410

Sutter A, Riopelle RJ, Harris-Warrick RM, Shooter EM (1979) NGF receptors: characterization of two distinct classes of binding sites on chick embryo sensory ganglia cells. J Biol Chem 254: 5972–5982

Taniuchi M, Schweitzer JB, Johnson EM (1986a) Phosphorylation of nerve growth factor receptor proteins in sympathetic neurons and PC12 cells. J Biol Chem 261: 13342–13349

Taniuchi M, Clark HB, Johnson EM (1986b) Induction of nerve growth factor receptor in Schwann cells after axotomy. Proc Natl Acad Sci USA 83: 4094–4098

Thorpe LW, Stach RW, Hashim GA, Marchetti D, Perez-Polo JR (1987) Receptors for nerve growth factor on spleen mononuclear cells. J Neurosci Res 17: 128–134

Upton C, deLange A, McFadden G (1987) Tumorigenic pox viruses: Genomic organization and DNA sequence of the telomeric region of the Shope fibroma virus. Virology 160: 20

Yamamoto T, Bishop RW, Brown MS, Goldstein JL, Russell DW (1986) Deletion in cysteine-rich region of LDL receptor impedes transport to cell surface in WHHL rabbit. Science 232: 1230–1237

Tissue Localization of Nerve Growth Factor and Nerve Growth Factor Receptors*

M. BOTHWELL[1]

1 Introduction

The pioneering investigations of LEVI-MONTALCINI and coworkers established that nerve growth factor (NGF) is an important physiological regulator of neurons of the peripheral nervous system. No doubt in part because of the dramatic and specific effects on the sympathetic and sensory nervous system resulting from immunochemically depriving rodents of NGF (LEVI-MONTALCINI and BOOKER 1960; JOHNSON et al. 1983), the notion that the action of NGF was directed exclusively to these peripheral neurons became firmly entrenched. A functional model which proposed that NGF was produced by targets of sensory and sympathetic innervation, regulating the specificity of axon growth to the target and the survival of neurons which found the correct target, became dogmatically accepted. Over the last few years it has become clear that these

* This work was supported by US Public Health Service grants NS-23343 and HL-43397
[1] Department of Physiology & Biophysics, School of Medicine, University of Washington, Seattle, WA 98195, USA

Current Topics in Microbiology and Immunology, Vol. 165

concepts are only partially correct. Aided in substantial part by improved methods for characterization of sites of expression of NGF and its receptor, it has been shown that the biology of NGF action is much more complicated than was initially imagined.

2 Potent Sources of NGF

Efforts to characterize the biologically important sites of NGF synthesis have been complicated by the existence of peculiar species-specific sites of abundant NGF synthesis. As discussed by Fahnestock in this volume, NGF is produced in substantial quantities in the salivary glands of mice and the African rat *Mastomys*, in the prostate gland of guinea pigs, and in bovine seminal vesicles, while no high potency source of NGF has been found in humans (HEINRICH and MEYER 1988). Whether these sources of NGF have specific biological roles in the species where they occur has not been established. What is clear, however, is that the NGF which supports the development of the peripheral nervous system is derived from other, less potent sources. This is implicit in the observation that many vertebrate species such as humans do not possess tissues where NGF is stored in large concentrations and that surgical removal of mouse salivary glands does not effect development of the peripheral nervous system.

While high potency sources of NGF have confused efforts to understand the physiology of NGF action, they have benefited efforts to characterize the factor biochemically, as described by Fahnestock in this volume. NGF exists in various oligomeric forms, the distribution of which in various biological sources has not been fully worked out. Nonetheless, the biologically active component consists of homodimeric β-NGF, hereafter in this chapter referred to simply as NGF. The availability of high abundance sources of NGF has also been invaluable in permitting the generation of specific antibodies and molecular clones of NGF. Application of these molecular probes has been essential for characterizing biologically relevant sources of NGF, as described below.

3 Techniques for Identification of Biological Sources of NGF

The early literature concerning localization of NGF in various tissues in many cases is best ignored. Some of the earliest studies employed biological assays with misleading results in some instances because the existence of other factors with NGF-like biological activity, such as fibroblast growth factor (FGF) and brain-derived neurotropic factor (BDNF), was not recognized. However, in cases where it can be demonstrated that biological activity is eliminated by specific antibodies to NGF, biological assays can be quite helpful.

A second source of confusion in the early literature resulted from the use of inappropriate immunoassays. Tissues and blood serum contain one or more NGF binding proteins (SUDA et al. 1978), including α-2 macroglobulin (KOO and STACH 1989), which interfere in conventional immunoassays giving false positive or false negative results depending on the assay used. Not until a two-site immunoassay for NGF was developed (SUDA et al. 1978; KORSCHING and THOENEN 1983) did it become possible to meaningfully assay NGF levels in tissue.

Interpretation of results of NGF localization must take into account one additional confusing factor. NGF-responsive neurons have the capacity to take up NGF at axon endings and transport it retrogradely to the cell body where it may accumulate in intact form in substantial quantities (STOEKEL and THOENEN 1975). Thus in some cases, localization of NGF protein may identify targets of NGF action rather than sites of NGF biosynthesis. In this regard, development of hybridization probes for NGF mRNA has been invaluable as this permits one to distinguish between sites of synthesis and sites of accumulation of NGF.

4 Techniques for Identification of Cellular Sites of NGF Receptor Expression

For many years, the only available method for assay of NGF receptor (NGFR) was binding of ^{125}I-NGF. Used on dissociated cells or cell membranes, this approach has led to the characterization of high and low affinity subclasses of NGFR which appear to be two functionally different forms of the same receptor protein, as described by CHAO in this volume. Such in vitro binding assays, while quantitative, are poorly suited for identification of the specific cell types within a tissue which express NGFR. Other techniques which similarly combine excellent quantitation but poor cellular resolution include northern blot or nuclease protection assays for NGFR mRNA and electrophoretic analysis of NGFR labeled by affinity cross-linking of ^{125}I-NGF, with or without subsequent immunoprecipitation. Where identification of NGFR positive cell types is more important than precise quantitation, in situ techniques such as autoradiographic localization of ^{125}I-NGF to tissue sections, in situ hybridization to NGFR mRNA, and immunohistochemical localization of receptors with specific antibodies are more appropriate. In this review I will discuss primarily reports employing the latter class of techniques. In comparing results obtained with these techniques it is important to remember that ^{125}I-NGF binding labels preferentially receptors of the high affinity class. In contrast, as high and low affinities derive from the same NGFR mRNA, in situ hybridization does not discriminate between high and low affinity receptor expression. It is not clear whether immunohistochemical localization of receptors preferentially labels one affinity class of receptors. It has been suggested that the commonly used antibody to rate NGFR 192 IgG (CHANDLER et al. 1984), recognizes only the low affinity class of receptors (YAN et al. 1988). However, some results suggest that the antibody recognizes both receptor classes (GREEN and GREENE 1986). Antibodies against human NGFR (ROSS et al. 1984; MARANO et al. 1987) appear to recognize both high and low affinity receptors as these antibodies block biological actions of NGF thought to be mediated by the high affinity form of receptor (HEMPSTEAD et al. 1989).

5 NGFR in the Peripheral Nervous System

The results of immunodeprivation of NGF in neonatal or prenatal animals, coupled with studies of responses in tissue culture, indicate that NGF supports the survival and differentiation of sympathetic neurons and the nociceptive subclass of neural crest-derived primary sensory neurons. By contrast, parasympathetic neurons, placodally derived sensory neurons, and proprioceptive, neural crest-derived, sensory neurons apparently do not require NGF for survival (PEARSON et al. 1983). The pattern of expression of NGFR in the peripheral nervous system is not wholly concordant with the subpopulations of neurons known to be dependent on NGF. NGFR expression on neurons of the peripheral nervous system has been assessed in vivo at the cellular level by binding of ^{125}I-NGF to tissue sections of chick and rat (RAIVICH et al. 1985; RAIVICH et al. 1987; RAIVICH and KREUTZBERG 1987; RICHARDSON et al. 1986), by immunohistochemical techniques in rat and primates (YAN and JOHNSON 1988; MARANO et al. 1987; SCHATTEMAN et al. 1986; KOH et al. 1989) and by in situ hybridization to chick and rat tissue sections (ERNFORS et al. 1989; GOEDERT et al. 1989; PERUSIO and BROOKS 1988; HEUER et al. 1990; KOH et al. 1989). By all of these techniques, NGFR expression is observed in placode-derived sensory neurons as well as in neural crest-derived sensory neurons (RAIVICH et al. 1987; YAN and JOHNSON 1988; HEUER et al. 1990). While ^{125}I-NGF binding studies in chicken suggested that epibranchial placode-derived, cranial sensory neurons express NGFR only transiently and only at low levels (RAIVICH et al. 1987), in situ hybridization revealed that epibranchial placode-derived neurons express NGFR mRNA in the same quantities and in the same temporal pattern as neural crest-derived neurons (HEUER et al. 1990). In agreement with the latter finding, in rat, NGFR immunoreactivity is expressed in placode-derived sensory neurons at similar levels and in a similar development temporal pattern as for neural crest-derived sensory neurons (YAN and JOHNSON 1988). Although one recent study suggested that NGF binding activity is not expressed in sensory neurons until after sensory axons have reached their targets (DAVIES et al. 1987), in situ hybridization revealed that NGFR mRNA is expressed, albeit at low levels, from the earliest stages of ganglionic development (HEUER et al. 1990) and that NGFR immunoreactivity is present in rat and chick neurons from the earliest time of formation of dorsal root ganglia (YAN and JOHNSON 1988; HEUER et al. 1990).

Indeed, premigratory cells of the neural crest and epibranchial placode express NGFR mRNA (HEUER et al. 1990). In apparent contradiction to this finding, the high affinity form of NGFR has not been detected on premigratory neural crest cells in culture (BERND 1985). Possible explanations for these differing results are that expression of NGFR mRNA may precede expression of cell surface receptors by a significant length of time; that in situ hybridization may be a more sensitive method of detecting low levels of NGFR expression; or that neural crest cells express only the low affinity form of receptor. Early in development, NGFR are expressed at uniform levels in the majority of neural crest and epibranchial-derived sensory neurons, but as neurons become fully differentiated there is a heterogeneous distribution of receptor expression. Most or all sensory neurons express NGFR at a detectable level, but a subpopulation of cells are much more strongly positive than the remainder (RICHARDSON et al. 1986; RAIVICH et al. 1985; ERNFORS et al. 1989; VERGE et al. 1989; HEUER et al. 1990). It is probable that the strongly positive subpopulation corresponds to the nociceptive neurons. This notion is supported by the observation that NGFR immunoreactivity is detected in spinal cord axonal projections in the substantia gelatinosa where nociceptive projections predominate but not in dorsal horn laminae receiving proprioceptive projections (SCHATTEMAN et al., manuscript in preparation). Thus, NGFR expression is highest in the neuronal subpopulation which is dependent on NGF for survival.

While NGF-dependent neuronal survival is characteristic of sympathetic neurons but not parasympathetic neurons, NGFR is expressed in both. YAN and JOHNSON (1988) reported that rat sympathetic neurons show only modest levels of NGFR immunoreactivity, irrespective of the stage of development, while parasympathetic and enteric neurons are only transiently positive. BUCK et al. (1987) reported that NGFR mRNA levels in rat sympathetic ganglia rise several fold postnatally. In rats, by in situ hybridization as well as by immunohistochemical techniques, investigators in my laboratory have observed substantial and comparable levels of NGFR expression in both sympathetic and some parasympathetic neurons throughout development and into adulthood (SCHATTEMAN et al. 1986; SCHATTEMAN et al., unpublished). We also have observed substantial expression of NGFR mRNA in enteric neurons in chick embryos (HEUER et al. 1990) and in both fetal and adult human enteric neurons (SCHATTEMAN, unpublished). In adult human tissue, only a subfraction of enteric neurons are NGFR positive. The possible significance of expression of NGFR in cell types not apparently responsive to NGF is discussed below.

In addition to neurons of the peripheral nervous system, glial cells also express NGFR. Both Schwann cells and perineurial fibroblasts of peripheral nerves express NGFR (ZIMMERMAN and SUTTER 1983; ROSS et al. 1984). During embryogenesis, NGFR is expressed on Schwann cells of all peripheral nerves, while in adulthood NGFR is expressed on all Schwann cells of sympathetic and parasympathetic nerves and on nonmyelinating Schwann cells of small peripheral sensory nerves, but not on Schwann cells of pure motor nerves or large mixed nerves such as the sciatic nerve (YAN and JOHNSON 1988; SCHATTEMAN et al. 1988; CHESA et al. 1988; SCARPINI et al. 1988; THOMPSON et al. 1989; SCHATTEMAN, unpublished work). Various cell types related in lineage to Schwann cells or perineurial fibroblasts also express NGFR. These include capsular cells of Meissner, Pacianian, and Ruffini sensory corpuscles (SCHATTEMAN et al., manuscript in preparation), ganglionic satellite cells (ZIMMERMAN and SUTTER 1983; MARANO et al. 1987), and satellite cells of adrenal medulla and carotid body (THOMPSON et al. 1989). Nerve sheath tumors universally express NGFR (THOMPSON et al. 1989). JOHNSON and coworkers have argued that Schwann cells are not capable of responding to NGF and suggest that Schwann cell NGFR may function to bind NGF and "present" it to axons of responsive neurons (TANIUCHI et al. 1986). However, SEILHEIMER and SCHACHNER (1986) have reported that Schwann cells *are* biologically responsive to NGF, inducing expression of the adhesive protein L1. In addition, as Schwann cells themselves produce NGF (see below) it must be considered that they may be self-stimulated in an autocrine fashion, minimizing effects of added NGF.

6 Sources of NGF for the Peripheral Nervous System

For many years a central hypothesis of workers in the NGF field has been that sensory and sympathetic neurons derive NGF trophic support from the targets which they innervate. Consistent with this hypothesis, there is good correlation between density of sympathetic innervation of various tissues and levels of NGF and NGF mRNA in those tissues (KORSCHING and THOENEN 1983; SHELTON and REICHARDT 1984; HEUMANN et al. 1984; KORSCHING and THOENEN 1988). In a target of dense sensory innervation, the maxillary process of mice, NGF mRNA is expressed in the epithelium and, to a lesser extent, the adjacent mesenchyme. (DAVIES et al. 1987). Remarkably, NGF does not appear in this target during embryogenesis until *after* sensory axons first arrive, implying that target-derived NGF cannot play any role in directing axons to their targets but many function solely in providing trophic support to neurons after innervation has been accomplished. It must be noted that the axons of primary sensory neurons bifurcate, sending branches into the spinal cord or brain stem as well as branches to the peripheral target. As noted by RAIVICH et al. (1985), the presence of abundant NGFR on the central projections of sensory neurons suggests that the CNS might also provide NGF trophic support. Central rhizotomy of dorsal root ganglia demonstrates that sensory neurons do receive trophic support from their central projections (YIP and JOHNSON 1984). The levels of NGF mRNA in spinal cord are low but not zero, and substantially higher levels of NGF mRNA are present in brain stem in regions receiving central projections from cranial sensory ganglia (SHELTON and REICHARDT 1986; LARGE et al. 1986). It is interesting to speculate that the apparent resistance of some cranial sensory ganglia to the effects of immune NGF deprivation might result in part from their access to an

NGF supply in brainstem which would not be accessible to circulating antibodies.

The production of NGF appears to be a rather common property of epithelial cells. Analogous to the relationship of sensory nerve fibres to cutaneous epithelium, the acoustic neuroepithelial hair cells of the cochlear organ of Corti are innervated by axons of the sensory neurons of the cochlear ganglion. Thus, perhaps it is not surprising that these neuroepithelial cells contain NGF (DEPRES et al. 1988). However, it is more difficult to envision a neural function of the NGF produced by epithelial cells of testes and epididymis. Thus it has been suggested that NGF acts on spermatozoa (AYER-LELIEVRE et al. 1988a). Another site of NGF production of uncertain functional significance is the pituitary gland (LAHTINEN et al. 1989; SOINILA et al. 1988).

In mouse salivary gland and heart ventricle, sites of sympathetic innervation, NGF immunoreactivity appears embryologically at the time of ingrowth of sympathetic axons and not before. Thus, for sympathetic neurons as for sensory neurons, it appears that NGF functions as a trophic agent but is not involved in chemotropic direction of axons to targets (KORSCHING and THOENEN 1988; CLEGG et al. 1989).

In addition to targets of innervation, another important source of NGF for the peripheral nervous system is the nerve itself. Schwann cells of peripheral nerves contain NGF immunoreactivity (RUSH 1984). Following axotomy of peripheral nerves, nonneuronal cells associated with the nerve (Schwann cells and perhaps also perineurial fibroblasts) are induced to produce NGF by macrophage-derived interleukin-1 (IL-1) (HEUMANN et al. 1987; LINDHOLM et al. 1987, 1988). This almost certainly has a significant role in promoting nerve regeneration. RUSH (1984) suggested that Schwann cells may be a major source of NGF for growning axons during embryonic development but THOENEN and coworkers (DAVIES et al. 1987) concluded that embryonic production of NGF in Schwann cells is small compared to NGF production by the target. The mechanism of induction of NGF expression in injured nerves is discussed below.

7 NGFR in the Central Nervous System

Apart from the NGFR on central projections of sensory neurons, in the adult central nervous system NGFR is expressed in high levels only in neurons of the cholinergic magnocellular complex of basal forebrain (HEFTI et al. 1986; SPRINGER et al. 1987; SCHATTEMAN et al. 1988; KORDOWER et al. 1988; YAN and JOHNSON et al. 1988; DAWBARN et al. 1988a, b; DREYFUS et al. 1989; HEFTI and MASH 1989; BATCHELOR et al. 1989; MUFSON et al. 1988). However, lower levels of NGFR in adult rat brain have been described in Purkinje cells of cerebellum (PIORO and CUELLO 1988) and in neurons of caudate-putamen, ventral premammillary nucleus, mesencephalic trigeminal nucleus, prepositus hypoglossal nucleus, raphe nucleus, and nucleus ambiguus (KOH et al. 1989). In my laboratory we have observed NGFR immunoreactivity in adult primate brains in the basal forebrain cholinergic magnocellular complex, neurons of caudate-putamen, mesencephalic trigeminal nucleus, and nucleus ambiguous (SCHATTEMAN et al. 1988; SCHATTEMAN et al., unpublished work). The biological role of NGF in basal forebrain cholinergic neurons is described in detail by GAGE et al. in this volume. In contrast to the limited expression of NGFR in adult CNS neurons, during fetal and early postnatal development NGFR is transiently expressed in a remarkable variety of neuronal populations. In chicken, according to restults of NGF blinding and mRNA in situ hybridization and solution hybridization, NGFR is expressed at substantial levels in every brain region embryologically (HEUER et al. 1990; BUCK et al. 1987; RAIVICH et al. 1987; LARGE et al. 1989). The periventricular germinal neuroepithelium is generally devoid of NGFR expression, while a uniform low level of expression is present throughout the marginal zone. Much higher levels of expression appear in specific nuclear groups as they differentiate (RAIVICH et al. 1987; ERNFORS et al. 1989; HEUER et al. 1990). Neuronal cell types which express NGFR during late embryogenesis include spinal cord lateral motor column and various brain stem motor nuclei, mesencephalic trigeminal sensory nucleus, isthmal nuclei, epithalamus, cerebellar Purkinje cells and cells of the external granular layer, and deep cerebellar nuclei (ERNFORS et al. 1989; HEUER et al. 1990; VON BARTHELD, HEUER and BOTHWELL, unpublished work). In rats at comparable stages of development, the generalized expression of NGFR mRNA in the marginal zone is not observed but NGFR hybridization appears in specific groups of differentiated neurons generally corresponding in identity and level of expression to that observed in chicken (E. WHEELER, unpublished results). NGFR immunoreactivity in developing rat has been localized to lateral geniculate nucleus, medial terminal nucleus of the accessory optic tract, cortical subplate cells, olivary pretectal nucleus, ventral and dorsal cochlear nucleus, superior olive, nucleus of the lateral lemniscus, cuneate nucleus, gracile nucleus, ventroposterior thalamic nucleus, amygdala, medial nucleus of inferior olive, and cerebellar Purkinje cells, deep nuclei, and external granule layer (YAN and JOHNSON 1988; ECKENSTEIN 1988; ALLENDOERFER et al. 1990). Retina also expresses NGFR on several cell types. Early in retinal development, both retinal ganglion cells and cells of the inner nuclear layer express NGFR in rat an in chick (HEUER et al. 1990; ECKENSTEIN 1988). NGFR positive cells of the inner nuclear layer include amacrine cells (S. PATTERSON, unpublished work; C. VON BARTHELD, manuscript in preparation). Expression in most of these sites decreases substantially in late prenatal or early postnatal development. In human and macaque embryos, transient expression has been observed on inferior olive and climbing fiber projections to cerebellum, on cerebellar Purkinje cells, and external granule layer, and on deep cerebellar nuclei (SCHATTEMAN et al. 1988; E. SCHATTEMAN, unpublished work).

NGFR immunoreactivity is also present, in rat and primates, on specific CNS glial populations, including glia of olfactory glomeruli (VICKLAND et al. 1989),

Müller glia of adult retina, astrocytes of embryonic retina and optic nerve (SCHATTEMAN et al. 1988; E. SCHATTEMAN and S. PATTERSON, unpublished work), and ventricular subependymal cells and tanycytes (KOH et al. 1989).

Of these novel sites of NGFR expression, perhaps the most remarkable is that of motor neurons. As shown by ^{125}I-NGF autoradiography, receptor immunohistochemistry, and receptor mRNA in situ hybridization, motor neurons in spinal cord lateral motor column and various brain stem motor nuclei transiently express high levels of NGFR. In spinal cord motor neurons, the timing of NGFR expression correlates approximately with the time of normal developmental cell death. By adulthood, NGFR levels are near the limit of sensitivity of immunohistochemical and in situ hybridization detection.

It must be noted that for most of these neuronal populations, there is, as yet, no direct evidence for biological regulation by NGF. Three exceptions are basal forebrain cholinergic neurons, for which NGF promotes survival and cholinergic phenotype (see this volume); embryonic trigeminal motor neurons, for which NGF stimulates neurite outgrowth in vitro (HEATON 1987); and axotomized retinal ganglion cells, for which NGF promotes survival in rats (CARMIGNOTO et al. 1989) and regeneration of axons in amphibians (TURNER and GLAZE 1977).

As to why NGFR is expressed on cells which apparently do not respond to NGF, several possible explanations exist. The first is that expression of NGFR on these cells is functionally irrelevant; I believe this is unlikely. A second possibility is that NGF does not modulate readily apparent cellular properties such as cell morphology or cell survival and growth but rather specific biochemical properties of cells (for example, synthesis of particular neurotransmitters or neuropeptides). Such regulation can only be detected by appropriate specific assays. A third possibility, especially in culture, is that enough NGF is produced by cells in the culture to mask the effects of added NGF. A fourth possibility is that NGF acts in concert with another factor so that NGF alone is necessary but no sufficient to give a response. Thus, identification of the regulatory role of NGF for such cells will challenge the resourcefulness and imagination of the investigator.

8 Sources of NGF for Central Neurons

Most extensively studied has been the source of NGF for basal forebrain magnocellular neurons. The cholinergic neurons of medial septal nucleus project to hippocampus, which has the highest level of NGF mRNA of any adult brain region (SHELTON and REICHARDT 1986; KORSCHING et al. 1985; WHITTEMORE et al. 1988). Numerous reports support the belief that NGF is produced in hippocampus, taken up by cholinergic axons, and transported retrogradely back to septal cell bodies. The results of in situ hybridization suggest that the site of NGF biosynthesis in hippocampus is neuronal; NGF mRNA is detected in hippocampal pyramidal and granule cells (WHITTEMORE et al. 1986; GALL and ISACKSON 1989) and these cells immunostain for NGF (AYER-LELIEVER 1988b). The chapter by GAGE et al. in this volume, provides additional discussion of this topic.

Embryologically, NGF mRNA appears transiently in cerebellum with a time course which correlates excellently with that of NGFR expression on neurons in this region. NGF mRNA also is expressed transiently in high levels in rat fetal olfactory bulb (LU et al. 1989). While abundant expression of NGFR mRNA in olfactory bulb during this period was demonstrated, our results suggest that this mRNA derives predominantly from meninges rather than from neurons (HEUER et al. 1990; E. WHEELER, unpublished work). NGFR expression on neurons intrinsic to olfactory bulb and on sensory neuroepithelial cells of the olfactory mucosa does not appear until postnatal periods (S. PATTERSON, unpublished work). In chicken, NGF mRNA is present in high levels in the eye during the period of NGFR expression on retinal neurons (EBENDAL et al. 1986). In rat, NGF immunoreactivity is present in retinal ganglion cells (AYER-LELIEVRE et al. 1983). As discussed above, it is not clear whether these cells make NGF or accumulate it by axonal uptake and retrograde transport. If these cells produce NGF, then, as ganglion cells also express NGFR, this suggests the possibility of an autocrine role for NGF in retinal ganglion cells.

Also, it must be considered that an important source of NGF for CNS neurons may be external to the CNS proper. NGF is present in CSF of early postnatal rats, apparently deriving at least in part from the choroid plexus (PATTERSON and BOTHWELL 1989). As NGF artificially introduced into CSF has been shown to easily gain access to and support basal forebrain cholinergic neurons (see GAGE et al., this volume), the NGF endogenously produced by choroid plexus may be biologically important.

Astrocytes produce NGF in tissue culture (LINDSAY 1979) as do microglia (MALLAT et al. 1989) raising the possibility that glia may be an important source of NGF in vivo. As yet there is little direct evidence that this is the case.

9 NGF and NGFR Expression in Nonneural Tissue

NGFR is expressed in a wide variety of cell types seemingly unrelated to the nervous system, suggesting that NGF may have wide-ranging biological roles outside the nervous system. The pioneering study of RAIVICH et al. (1985) demonstrated expression of high affinity NGF binding sites on embryonic mesenchyme and developing muscle. Subsequent studies have confirmed and extended these results by immunohistochemical localization of NGFR and in situ localization of NGFR mRNA (YAN and JOHNSON 1988; ERNFORS et al. 1988; HEUER et al. 1990). In embryos, NGFR is expressed in somites and, as somites become subcompartmentalized, in sclerotome and dermatome. The mesenchyme of the branchial arch and of early limb buds is strongly NGFR positive. During subsequent development, NGFR disappears from these mesenchymal cells and then reappears on myoblasts just prior to the time of myoblast fusion. NGFR also is expressed transiently on the mesenchyme surrounding developing epithelial structures, such as hair follicles, salivary glands, and lung bronchioles, on embryonic perivascular cells, and on embryonic meninges. NGFR is not expressed on most epithelial structures during embryogenesis; exceptions where NGFR is expressed transiently include the chick and rat auditory vesicle epithelium (BERND and REPRESSA 1989; VON BARTHELD et al. submitted), dental epithelium (YAN and JOHNSON 1988; BYERS et al. (1990)) and the luminal epithelium of fetal human and rat intestine (SCHATTEMAN, unpublished work; E. WHEELER, unpublished). In adult tissue, mesenchymal tissues generally are devoid of NGFR with the exception of perineurial fibroblasts, some perivascular cells, and dental pulp cells (ROSS et al. 1984; CHESA et al. 1988; THOMPSON et al. 1989; BYERS et al. 1990). However, NGFR expression is induced in a manner reminiscent of the embryonic pattern of expression in the stromal tissue of tumors such as carcinomas (CHESA et al. 1988; PERUSIO and BROOKS 1988; THOMPSON et al. 1989). A variety of adult basal epithelial cell types express NGFR including glandular myoepithelium, hair follicle outer root sheath cells, salivary gland ducts, lung bronchioles, and oral mucosal epithelium (CHESA et al. 1988; THOMPSON et al. 1988, 1989; S. PATTERSON, unpublished work). The only one of these nonneural sites of NGFR expression where NGF action has been carefully investigated is embryonic auditory vesicle epithelium, for which NGF is a mitogen (REPRESSA and BERND 1990).

10 NGF Action in Hematopoietic Cells

NGF stimulates mast cell proliferation (ALOE 1988) and promotes mast cell degranulation (TOMIOKA et al. 1988). However, the NGFR on these cells have not been fully characterized. NGF has been reported to act as a mitogen for T and B lymphocytes, and NGF receptors on these cells have been described (THORPE and PEREZ-POLO 1987; MORGAN et al. 1989). MATSUDA et al. (1988) have reported that NGF stimulates granulocyte differentiation in peripheral blood cell culture, possibly by stimulating production of granulocyte colony stimulating factor (G-CSF) by T lymphocytes. However, we have failed to detect NGFR immunoreactivity on splenic or lymph node lymphocytes (THOMPSON et al. 1989) suggesting that, if NGFR is expressed by lymphocytes, receptor levels must be very low or, alternatively, present only on a small proportion of lymphocytes. CHESA et al. (1988) reported that NGFR immunoreactivity is present on lymphocytes of lymph node germinal centers. However, our own work indicates that this germinal center immunoreactivity derives primarily, if not exclusively, from follicular dendritic cells, rather than lymphocytes (THOMPSON et al. 1989). We have failed to observe any mitogenic effect of NGF on rat lymph node lymphocytes (unpublished).

11 Regulated Expression of NGF and NGFR in Neural Tissue

Expression of NGFR in neural tissue is regulated by complex mechanisms. In dorsal root ganglion sensory neurons (VERGE et al. 1989) and magnocellular cholinergic neurons of basal forebrain (see GAGE et al., this volume) NGF appears to upregulate expression of its own receptor. Axotomy of these cells causes diminshed expression of NGFR. This is apparently due to isolation from NGF derived from axon terminal fields since exogenously supplied NGF returns NGFR levels to normal. However, adult spinal cord motor neurons cells initially lacking NGFR express the receptor after axotomy (ERNFORS et al. 1989). The mechanism of this induction is not clear. It is tempting to speculate that re-expression of NGFR after injury may be a common feature among the numerous neuronal cell types which normally express NGFR late in development but not in adulthood.

NGFR expression on Schwann cells of large peripheral nerves (sciatic nerve, sural nerve) is induced after nerve injuries such as experimental axotomy or neurodegenerative disease (TANIUCHI et al. 1986; HEUMANN et al. 1987; SOBUE et al. 1988). Neither the mechanism of induction nor its functional significance is clear. Schwann cell NGFR expression is repressed by contact with axons in a mechanism which is not cAMP dependent (LEMKE and CHAO 1988; but see MOKUNO et al. 1988).

It is important to note that the same types of injuries which induce Schwann cells to express NGFR also induce expression of NGF. NGF expression in Schwann cells after nerve injury occurs in two temporal phases. Following a rapid initial increase in NGF synthesis, the mechanism of induction of which is unclear, a second larger burst of NGF synthesis occurs which is apparently due to the effects of IL-1 released by infiltrating macrophages (HEUMANN et al. 1987; LINDHOLM et al. 1988). Also, macrophages themsleves may be a source of NGF as lipopolysaccharide-stimulated brain macrophages produce NGF (MALLAT et al. 1989).

Numerous recent reports demonstrate that brain injury of diverse types can stimulate NGF production. Types of injury which are effective include surgical wounds (WHITTEMORE et al. 1987; GASSER et al. 1986), hypoxia (LOREZ et al. 1989), and limbic seizure (GALL and ISACKSON 1989). In limbic seizure, the source of induced NGF production is neuronal. In other types of injury the cell type responsible has not been identified, but it has been suggested that the source might be inflammatory cells such as reactive astrocytes (GASSER et al. 1986) or macrophages (MALLAT et al. 1989).

Regulation of NGF production has been extensively studied in tissue culture. IL-1 induces NGF production by fibroblasts (LINDHOLM et al. 1988), while NGF production is depressed by glucocorticoids (HOULGATE et al. 1989). Prostaglandins and β-adrenergic compounds induce NGF production in astrocytoma cells (DAL TOSO et al. 1988; FURAKAWA et al. 1989; MOCOHETTI et al. 1989; SCHWARTZ 1988). In marked contrast, norepinephrine *reduces* NGF production by smooth muscle cells of the iris (HELLWEGG et al. 1988).

12 Conclusions

The presence of NGF receptors on a wide variety of neuronal and nonneuronal cell types suggests that NGF may have a much wider range of regulatory roles than is currently recognized. It is also clear that a wide range of cell types have the capacity to produce NGF. Thus the relationship between various sites of production of NGF and various sites of action of NGF is likely to be extremely complex. The recent discovery that BDNF is structurally similar to NGF (Leibrock et al. 1989) raises the question whether BDNF might interact functionally with the NGF receptor and whether additional members of the NGF ligand family might remain to be discovered. One is left with the impression that we have hardly begun to grasp the significance of this system of biological regulators.

References

Allendoerfer KL, Shelton DL, Shooter EM, Shatz CJ (1990) Nerve growth factor receptor immunoreactivity is transiently associated with the subplate neurons of the mammalian cerebral cortex. Proc Natl Acad Sci USA 87: 187–190

Aloe L (1988) The effect of nerve growth factor and its antibody on mast cells in vivo. J Neuroimmunol 18: 1–12

Ayer-Le Lievre C, Ebendal T, Olson L, Seiger A (1983) Localization of nerve growth factor-like immunoreactivity in rat nervous tissue. Med Biol 61: 296–304

Ayer-LeLievre C, Olson L, Ebendal T, Hallbook F, Persson H (1988a) Nerve growth factor mRNA and protein in the testis and epididymis of mouse and rat. Proc Natl Acad Sci USA 85: 2628–2632

Ayer-LeLievre C, Olson L, Ebendal T, Seiger A, Persson H (1988b) Expression of the beta-nerve growth factor gene in hippocampal neurons. Science 240: 1339–1341

Batchelor PE, Armstrong DM, Blaker SM, Gage FH (1989) Nerve growth factor receptor and choline acetyltransferase colocalization in neurons within the rat forebrain: response to fimbria-fornix transection. J Comp Neurol 284: 187–204

Bernd P (1985) Appearance of nerve growth factor receptors on cultured neural crest cells. Dev Biol 112: 145–156

Bernd P, Represa J (1989) Characterization and localization of nerve growth factor receptors in the embryonic otic vesicle and cochleovestibular ganglion. Dev Biol 134: 11–20

Buck CR, Martinez HJ, Black IB, Chao MV (1987) Developmentally regulated expression of the nerve growth factor receptor gene in the periphery and brain. Proc Natl Acad Sci USA 84: 3060–3063

Byers MR, Schatteman GC, Bothwell M (1990) Multiple functions of nerve growth factor receptor in developing, aging and injured rat teeth. Development 109: 461–471

Carmignoto G, Maffei L, Candeo P, Canella R, Comelli C (1989) Effect of NGF on the survival of rat retinal ganglion cells following optic nerve section. J Neurosci 9: 1263–1272

Chandler CE, Parsons LM, Hosang M, Shooter EM (1984) A monoclonal antibody modulates the interaction of nerve growth factor with PC12 cells. J Biol Chem 259: 6882–6889

Chesa PG, Wolfgang JR, Thomson TM, Old LJ, Melamed MR (1988) Immunohistochemical analysis of nerve growth factor receptor expression in normal and malignant human tissues. J Histochem Cytochem 36: 383–389

Clegg DO, Large TH, Bodary SC, Reichardt LF (1989) Regulation of nerve growth factor mRNA levels in developing rat heart ventricle is not altered by sympathectomy. Dev Biol 134: 30–7

Dal Toso R, De Bernardi MA, Brooker G, Costa E, Mocchetti I (1988) Beta adrenergic and prostaglandin receptor activation increases nerve growth factor mRNA content in C6-2B rat astrocytoma cells. J Pharmacol Exp Ther 246: 1190–1193

Davies AM, Bandtlow C, Heumann R, Korsching S, Rohrer H, Thoenen H (1987) Timing and site of nerve growth factor synthesis in developing skin in relation to innervation and expression of the receptor. Nature 326: 353–357

Dawbarn D, Allen SJ, Semenenko FM (1988a) Immunohistochemical localization of beta-nerve growth factor receptors in the forebrain of the rat. Brain Res 440: 185–189

Dawbarn D, Allen SJ, Semenenko FM (1988b) Coexistence of choline acetyltransferase and nerve growth factor receptors in the rat basal forebrain. Neurosci Lett 94: 138–144

Despres G, Giry N, Romand R (1988) Immunohistochemical localization of nerve growth factor-like protein in the organ of Corti of the developing rat. Neurosci Lett 85: 5–8

Dreyfus CF, Bernd P, Martinez HJ, Rubin SJ, Black IB (1989) GABAergic and cholinergic neurons exhibit high-affinity nerve growth factor binding in rat basal forebrain. Exp Neurol 104: 181–185

Ebendal T, Larhammar D, Persson H (1986) Structure and expression of the chicken beta nerve growth factor gene. EMBO J 5: 1483–1487

Eckenstein F (1988) Transient expression of NGF-receptor-like immunoreactivity in postnatal rat brain and spinal cord. Brain Res 446: 149–154

Ernfors P, Hallbook F, Ebendal T, Shooter EM, Radeka MJ, Misko TP, Persson H (1988) Developmental and regional expression of beta-nerve growth factor receptor mRNA in the chick and rat. Neuron 1: 983–996

Ernfors P, Henschen A, Olson L, Persson H (1989) Expression of nerve growth factor receptor mRNA is developmentally regulated and increased after axotomy in rat spinal cord motoneurons. Neuron 2: 1605–1613

Furukawa Y, Tomioka N, Sato W, Satoyoshi E (1989) Catecholamines increase nerve growth factor mRNA content in both mouse astroglial cells and fibroblast cells. FEBS Lett 247: 463–467

Gall CM, Isackson PJ (1989) Limbic seizures increase neuronal production of messenger RNA for nerve growth factor. Science 245: 758–761

Gasser UE, Weskamp G, Otten U, Dravid AR (1986) Time course of elevation of nerve growth factor (NGF) content in hippocampus and septum following lesions of septohippocampal pathway in rats. Brain Res 376: 351–356

Goedert M (1989) Molecular cloning of the chicken nerve growth factor gene: mRNA distribution in developing and adult tissues. Biochem Biophys Res Commun 141: 1116–1122

Green SH, Greene LA (1986) A single Mr 103,000 ^{125}I-beta-nerve growth factor affinity-labeled species represents both the low and high affinity forms of the nerve growth factor receptor. J Biol Chem 261: 15316–16326

Heaton MB (1987) Influence of nerve growth factor on chick trigeminal motor nuclei explants. J Comp Neurol 265: 362–366

Hefti F, Mash DC (1989) Localization of nerve growth factor receptors in the normal human brain and in Alzheimer's disease. Neurobiol Aging 10: 75–87

Hefti F, Hartikka J, Salvatierra WJ, Weiner WJ, Mash D (1986) Localization of nerve growth factor receptors in cholinergic neurons of the human basal forebrain. Neurosci Lett 69: 37–41

Heinrich G, Meyer TE (1988) Nerve growth factor (NGF) is present in human placenta and semen but undetectable in normal and Paget's disease blood: measurements with an anti-mouse-NGF enzyme immunoassay using a recombinant human NGF reference. Biochem Biophys Res Commun 155: 482–486

Hellweg R, Bandtlow CE, Heumann R, Korsching S (1988) Nerve growth factor synthesis in cultured rat iris: modulation by endogenous transmitter substances. Exp Cell Res 179: 18–30

Hempstead BL, Schlierer LS, Chao MV (1989) Expression of functional nerve growth factor receptors after gene transfer. Science 243: 373–375

Heuer JG, Fatemie-Nainie S, Wheeler EW, Bothwell M (1990) Structure and developmental expression of the chicken NGF receptor. Dev Biol 137: 287–304

Heumann R, Korsching S, Scott J, Thoenen H (1984) Relationship between levels of nerve growth factor (NGF) and its mRNA in sympathetic ganglia and peripheral target tissues. EMBO J 3: 3183–3189

Heumann R, Lindholm D, Bandtlow C, Meyer M, Radeke MJ, Misko TP, Shooter E, Thoenen H (1987) Differential regulation of mRNA encoding nerve growth factor and its receptor in rat sciatic nerve during development, degeneration, and regeneration: role of macrophages. Proc Natl Acad Sci USA 84: 8735–8739

Houlgate T, Wion D, Brachet P (1989) Levels of nerve growth factor secreted by rat primary fibroblasts and iris transplants are influenced by serum and glucocorticodis. Dev Brain Res 47: 171–179

Johnson EM Jr, Osborne PA, Rydel RE, Schmidt RE, Pearson J (1983) Characterization of the effects of autoimmune nerve growth factor deprivation in the developing guinea pig. Neuroscience 8: 631–642

Koo PH, Stach RW (1989) Interaction of nerve growth factor with murine alpha-macroglobulin. J Neurosci Res 22: 247–261

Koh S, Oyler GA, Higgins GA (1989) Localization of nerve growth factor receptor mRNA and protein in adult rat brain. Exp Neurol 106: 209–221

Kordower JH, Bartus RT, Bothwell M, Schatteman GC, Gash DM (1988) Nerve growth factor receptor immunoreactivity in the nonhuman primate (Cebus apella): distribution, morphology and colocalization with cholinergic enzymes. J Comp Neurol 277: 465–486

Korsching S, Thoenen H (1983) Nerve growth factor in sympathetic ganglia and corresponding target organs in the rat:correlation with density of innervation. Proc Natl Acad Sci USA 80: 3513–3516

Korsching S, Thoenen H (1988) Developmental changes of nerve growth factor levels in sympathetic ganglia and their target organs. Dev Biol 126: 40–46

Korsching S, Auburger G, Heumann R, Scott J, Thoenen H (1985) Levels of nerve growth factor and its mRNA in the central nervous system of the rat correlate with cholinergic innervation. EMBO J 4: 1389–1393

Lahtinen T, Soinila S, Lakshmanan J (1989) Biological demonstration of nerve growth factor in the rat pituitary gland. Neuroscience 30: 165–170

Large TH, Bodary SC, Clegg DO, Weskamp G, Otten U, Reichardt LF (1986) Nerve growth factor gene expression in the developing rat brain. Science 234: 352–355

Large TH, Weskamp G, Helder JC, Radeke MJ, Misko TO, Shooter EM, Reichardt L (1989) Structure and developmental expression of the nerve growth factor receptor in the chicken central nervous system. Neuron 2: 1123–1134

Leibrock J, Lottspeich F, Hohn A, Hofer M, Hengerer B, Masiakowski P, Thoenen H and Barde Y-A (1989) Molecular cloning and expression of brain-derived neurotrophic factor. Nature 341: 149–154

Lemke G, Chao MV (1988) Axons regulate Schwann cell expression of the major myelin and nerve growth factor receptor genes. Development 102: 499–504

Levi-Montalcini R, Booker B (1960) Destruction of the sympathetic ganglia in mammals by an antiserum to NGF. Proc Natl Acad Sci USA 46: 384–391

Lindholm D, Heumann R, Meyer M, Thoenen H (1987) Interleukin-1 regulates synthesis of nerve growth factor in non-neuronal cells of rat sciatic nerve. Nature 330: 658–659

Lindholm D, Heumann R, Hengerer B, Thoenen H (1988) Interleukin 1 increases stability and transcription of mRNA encoding nerve growth factor in cultured rat fibroblasts. J Biol Chem 263: 16348–16351

Lindsay RM (1979) Adult rat brain astrocytes support survival of both NGF-dependent and NGF-independent neurons. Nature 282: 80–82

Lorez H, Keller F, Ruess G, Otten U (1989) Nerve growth factor increases in adult rat brain after hypoxic injury. Neurosci Lett 10: 339–344

Lu B, Buck CR, Dreyfus CF, Black IB (1989) Expression of NGF and NGF receptor mRNAs in the developing brain: evidence for local delivery and action of NGF. Exp Neurol 104: 191–199

Mallat M, Houlgatte R, Brachet P, Prochiantz A (1989) Lipopolysaccharide-stimulated rat brain macrophages release NGF in vitro. Devi Biol 133: 309–311

Marano N, Dietzschold B, Earley JJ Jr, Schatteman GC, Thompson S, Grob P, Ross AH, Bothwell M, Atkinson BF, Koprowski H (1987) Purification and amino terminal sequencing of human melanoma nerve growth factor receptor. J Neurochem 48: 225–232

Matsuda H, Coughlin MD, Bienenstock J, Denburg JA (1988) Nerve growth factor promotes human hemopoietic colony growth and differentiation. Proc Natl Acad Sci USA 85: 6508–6512

Mocchetti I, De Bernardi MA, Szekely AM, Alho H (1989) Regulation of nerve growth factor biosynthesis by beta-adrenergic receptor activation in astrocytoma cells: a potential role of c-*fos* protein. Proc Natl Acad Sci USA 86: 3891–3895

Mokuno K, Sobue G, Reddy UR, Wurzer J (1988) Regulation of Schwann cell nerve growth factor receptor by cyclic adenosine 3′,5′-monophosphate. J Neurosci Res 21: 465–472

Morgan B, Thorpe LW, Marchetti D, Perez-Polo JR (1989) Expression of nerve growth factor receptors by human peripheral blood mononuclear cells. J Neurosci Res 23: 41–45

Mufson EJ, Bothwell M, Hersh LB, Kordower JH (1989) Nerve growth factor receptor immunoreactive profiles in the normal, aged human basal forebrain: colocalization with cholinergic neurons. J Comp Neurol 285: 196–217

Patterson SL, Bothwell M (1989) NGF in choroid plexus and CSF. Soc Neurosci Abstr 15: 868

Pearson J, Johnson EM, Brandler L (1983) Effects of antibodies to nerve growth factor on intrauterine development of derivatives of cranial neural crest and placode in the guinea pig. Dev Biol 96: 32–36

Pioro EP, Cuello AC (1988) Purkinje cells of adult rat cerebellum express nerve growth factor receptor immunoreactivity: light microscopic observations. Brain Res 455: 182–186

Raivich G, Kreutzberg GW (1987) The localization and distribution of high affinity beta-nerve growth factor binding sites in the central nervous system of the adult rat. A light microscopic autoradiographic study using (125I)beta-nerve growth factor. Neuroscience 20: 23–36

Raivich G, Zimmerman A, Sutter A (1985) The spatial and temporal pattern of beta NGF receptor expression in the developing chick embryo. EMBO J 4: 637–644

Raivich G, Zimmerman A, Sutter A (1987) Nerve growth factor (NGF) receptor expression in chicken cranial development. J Comp Neurol 256: 229–245

Represa J, Brend P (1989) Nerve growth factor and serum differentially regulate development of the embryonic otic vesicle and cochleovestibular ganglion in vitro. Dev Biol 134: 21–29

Richardson PM, Riopelle RJ (1984) Uptake of nerve growth factor along peripheral and spinal axons of primary sensory neurons. J Neurosci 4: 1683–1689

Shelton DL, Reichardt LF (1984) Expression of the beta-nerve growth factor gene correlates with the density of sympathetic innervation of effector organs. Proc Natl Acad Sci USA 81: 7951–7955

Shelton DL, Reichardt LF (1986) Studies on the expression of the beta nerve growth factor (NGF) gene in the central nervous system: Level and regional distribution of NGF mRNA suggest that NGF functions as a trophic factor for several distinct populations of neurons. Proc Natl Acad Sci USA 83: 2714–2718

Soinila S, Lakshmanan J, Lahtinen T (1988) Nerve growth factor-like activity in the rat pituitary intermediate lobe. Neurosci Res 6: 123–130

Springer JE, Koh S, Tayrien MW, Loy R (1987) Basal forebrain magnocellular neurons stain for nerve growth factor receptor: correlation with cholinergic cell bodies and effects of axotomy. J Neurosci Res 17: 111–118

Stoekel K, Thoenen H (1975) Retrograde transport of nerve growth factor: specificity and biological importance. Brain Res 85: 337–341

Suda K, Barde YA, Thoenen H (1978) Nerve growth factor in mouse and rat serum: Correlation between bioassay and radioimmunoassay determination. Proc Natl Acad Sci USA 75: 4042–4046

Taniuchi M, Clark HB, Johnson EM Jr (1986) Induction of nerve growth factor receptors in Schwann cells after axotomy. Proc Natl Acad USA 83: 4094–4098

Thomson TM, Rettig WJ, Chesa PG, Green SH, Mena AC, Old LJ (1988) Expression of human nerve growth factor receptor on cells derived from all three germ layers. Exp Cell Res 174: 533–539

Thompson S, Schatteman GC, Gown AM, Bothwell M (1989) A monoclonal antibody against nerve growth factor receptor: immunohistochemical analysis of normal and neoplastic human tissue. Am J Clin Path 92: 415–423

Thorpe LW, Perez-Polo JR (1987) The influence of nerve growth factor on the in vitro proliferative response of rat spleen lymphocytes. J Neurosci Res 18: 134–139

Tomioka M, Stead RH, Nielsen L, Coughlin MD, Bienenstock J (1988) Nerve growth factor enhances antigen and other secretagogue-induced histamine release from rat peritoneal mast cells in the absence of phosphatidylserine. J Allergy Clin Immunol 82: 599–607

Turner JE, Glaze KA (1977) Regenerative repair in the severed optic nerve of the newt: effect of nerve growth factor. Exp Neurol 57: 687–697

Verge VM, Riopelle RJ, Richardson PM (1989) Nerve growth factor receptors on normal and injured sensory neurons. J Neurosci 9: 914–922

Vickland H, Kott JN, Westrum LE, Patterson SL, Bothwell M (1989) Electron microscopy shows NGF receptor labelling of glial-like cells in olfactory bulb. Soc Neurosci Abstr 15: 868

Whittemore SR, Ebendal T, Larkfors L, Olson L, Seiger A, Stromberg I, Persson H (1986) Developmental and regional expression of beta nerve growth factor messenger RNA and protein in the rat central nervous system. Proc Natl Acad Sci USA 83: 817–821

Whittemore SR, Friedman PL, Larhammar D, Persson H, Ebendal T (1988) Rat beta-nerve growth factor sequence and site of synthesis in the adult hippocampus. J Neurosci Res 20: 403–410

Yan Q, Johnson EM (1988) An immunohistochemical study of the nerve growth factor receptor in developing rats. J Neurosci 8: 3481–3498

Yip HK, Johnson EM (1984) Developing dorsal root ganglionic neurons require trophic support from their central processes: evidence for a role of retrogradely transported nerve growth factor from the central nervous system to the periphery. Proc Natl Acad Sci USA 81: 6245–6249

Zimmerman A, Sutter A (1983) Beta nerve growth factor (beta-NGF) receptors on glial cells. Cell–cell interaction between neurones and Schwann cells in cultures of chick sensory ganglia. EMBO J 2: 879–885

Nerve Growth Factor Function in the Central Nervous System*

F. H. GAGE[1], M. H. TUSZYNSKI[1], K. S. CHEN[1], A. M. FAGAN[1], and G. A. HIGGINS[2]

1 Introduction

A complete understanding of NGF function in the CNS does not exist. Instead, there is a collection of observations based on experiments conducted in the PNS and CNS, although more definitive results have come from experiments in the PNS. The conclusion that NGF is a neurotropic factor in the PNS biased experimental design and interpretation of results when a function for NGF in the CNS was sought. An example of design and interpretation bias is the very

* The research presented in this manuscript was supported by NIA AGO 6988; NIA AGO 0353A (MHT); Office of Naval Research; State of California DHHS; the Pew Foundation; and the McKnight Foundation

[1] Department of Neurosciences M-024, University of California at San Diego, La Jolla, CA 92093, USA

[2] Gerontology Research Center, Laboratory of Biological Chemistry, 4940 Eastern Ave., Baltimore, MD 21224, USA

Current Topics in Microbiology and Immunology, Vol. 165

significant paper by SCHWAB et al. (1979), showing for the first time retrograde transport of NGF in cholinergic neurons. These authors entitled the manuscript "Nerve growth factor (NGF) in the CNS: absence of specific retrograde transport and tyrosine hydroxylase induction in locus coeruleus and substantia nigra", documenting the *expectation* of retrograde transport in catecholamine neurons of the CNS, as in the PNS. Subsequent experiments, reviewed here, support a neurotrophic role for NGF in the CNS, although in the *cholinergic* rather than catecholaminergic system of CNS neurons.

During development more neurons are generated than are needed in the mature CNS. Some neurons undergo cell death at times when their growing axons compete for target territories, reducing the final number of neurons to that found in the adult (COWAN et al. 1984). The extent of neuronal cell death during development can be affected by experimentally manipulating the target area of the developing neurons to add or subtract target tissue (LANDMESSER and PILAR 1978; HAMBURGER and OPPENHEIMER 1982). These observations have led several investigators to propose that developmental cell death and survival are regulated by proteins presumably supplied by the target territory, which have been called *neurotrophic factors* (NTF) (THOENEN and BARD 1980). "Trophic" refers to the ability of one tissue, cell, or protein to support and/or nourish another; thus a neurotrophic factor is a chemical or molecule that is made in any cell that supports the survival of, or nourishes, neurons. *Trophic* differs markedly from *tropic*; the latter refers to the influence of one cell or tissue on the direction of movement or outgrowth of another. Thus a *neurotropic factor* is a chemical or molecule that can influence the direction and/or growth of an axon.

The neurotrophic hypothesis postulates that: (1) adult CNS neurons are supported and regulated by their respective NTFs, (2) proper maintenance of these neurons depends on adequate supply and utilization of the NTFs in situ, (3) an interference with NTF support, or "neurotrophic deficit", will result in defective performance or even degeneration and death of the target neurons, and (4) such trophic deficits may be the basis of degenerative central nervous system diseases (e.g., Parkinson's, motor neuron, or Alzheimer's disease) or normal aging (APPEL 1981).

The neurotrophic hypothesis is supported in the CNS by: (1) "developmental neuronal death", in which the excessive number of neurons produced during development is decreased to accommodate the limited target cell number (COWAN et al. 1984); (2) "retrograde neuronal degeneration", in which axotomized neurons cut off from their innervation target and surrounding glial cells undergo degeneration or even death (PEARSON et al. 1983); and (3) "pathological neuronal death", where specific populations of neurons degenerate and die (APPEL 1981; BARTUS et al. 1982). Among several explanations currently being tested, one that is commonly put forth for such neuronal death-inducing situations is that neurons normally depend for their continued health upon NTFs, supplied by their target and associated glial cells, and that disruption in this trophic supply causes their death.

Nerve growth factor (NGF) is currently the best characterized NTF (BARDE et al. 1983; BERG 1984; THOENEN and BARDE 1980; ULLRICH et al. 1983). NGF supports the survival in vitro and in vivo of sensory and sympathetic neurons from the PNS (GUNDERSON and BARRETT 1980) (neurotrophic). Furthermore NGF can attract and guide developing and regenerating axons, whether provided to the neuron in a soluble or immobilized form (GUNDERSON 1980), and may even guide axons whose neurons do not require NGF for survival (COLLINS and DAWSON 1983) (neurotropic).

2 NGF Function on CNS Neurons In Vitro

NGF can support the survival of embryonic and adult cells, induce and influence the direction of fiber outgrowth, and induce the synthesis of specific enzymes in vitro. The biological actions of NGF have been well-studied in neurons of the PNS and have recently been examined in the CNS.

Sympathetic and dorsal root ganglion cells have been extensively studied in culture. Cell death is found to occur in these systems (COWAN et al. 1984), perhaps as a result of competition for a limited supply of peripherally derived trophic factors. Sympathetic ganglion neurons derived from embryonic animals require NGF for their survival (BERG 1984; CHUN and PATTERSON 1977a, and b; GREENE 1977a; LEVI-MONTALCINI and ANGELETTI 1963). NGF is also able to promote the survival of dorsal root ganglion cells (CRAIN 1975; CRAIN and PETERSON 1974; GREENE 1977b; LEVI-MONTALCINI and ANGELETTI 1963) and can induce neurite extension in these cells (GREENE 1977b; LEVI-MONTALCINI and ANGELETTI 1963). Experiments demonstrating that the direction of growth of the neurites from NGF-responsive ganglion cells in culture is influenced by a NGF concentration gradient provided evidence for a tropic effect of NGF (CAMPENOT 1977; CHARLWOOD et al. 1972; EBENDAL and JACOBSON 1977; GUNDERSON 1985; GUNDERSON and BARRETT 1979, 1980; LETOURNEAU 1978). CAMPENOT (1977) showed that NGF can influence the direction of neurite outgrowth and support the survival of cells via uptake by a cell's peripheral projections without involvement of the cell soma. Cocultures of sympathetic ganglia and NGF rich tissue results in the innervation of the tissue by fibers from the ganglion cells (BURNSTOCK 1974; CHAMLEY et al. 1973; JOHNSON et al. 1972). NGF can also increase the activity of specific enzymes in neurons. The application of NGF to rat superior cervical ganglion cells in vitro increases tyrosine hydroxylase and dopamine beta hydroxylase activity in these cells (MACDONNELL et al. 1977; MAX et al. 1978; OTTEN et al. 1977). NGF is also able to induce choline acetyltransferase (ChAT) activity in sympathetic neurons (see THOENEN and BARDE 1980 for a review).

There is increasing evidence for a role for NGF in the CNS. Unlike catecholaminergic neurons in the PNS, catecholaminergic neurons in the CNS are unresponsive to NGF (Dreyfus et al. 1980; Olson et al. 1979). In the CNS, *cholinergic* neurons are responsive to NGF. Wainer et al. (1986) and Martinez et al. (1987) demonstrated that cultures of central cholinergic neurons require NGF for their survival, and Hartikka and Hefti (1988) found that NGF affects the survival of cultures of low plating density but not of high plating density. Although cholinergic cell cultures of high plating density do not require exogenous NGF for their survival, NGF can stimulate fiber growth of septal cholinergic neurons (Hartikka and Hefti 1988). Additionally, the direction of axonal outgrowth from cultured septal cells can be influenced by NGF. Gahwiler et al. (1987) showed that NGF can influence the growth of septal axons to cocultured hippocampal slices. NGF also has been shown to increase ChAT activity in cultures of neonatal and fetal basal forebrain cells (Gnahn et al. 1983; Hartikka and Hefti 1988; Hatanaka and Tsukui 1986; Hefti et al. 1985; Honegger et al. 1986; Honegger and Lenoir 1982; Martinez et al. 1987) and striatal cells (Martinez et al. 1985). Antibodies to NGF can block the observed increase in ChAT activity in basal forebrain cultures (Hartikka and Hefti 1988; Hefti et al. 1985) and reduce the number of acetylcholinesterase (AChE) positive cells in low plating density cultures (Hartikka and Hefti 1988). Additionally, cultures of hippocampal cells, which are the targets of the cholinergic neurons in the septum, were shown to contain NGF-like activity that could be blocked by antibodies to NGF (Crutcher and Collins 1982).

NGF is also produced by cultures of Schwann cells in the PNS and astrocytes in the CNS (Lindsay 1979; Norrgren et al. 1980; Rudge et al. 1985). The conditioned media from cultures of astrocytes can support the survival of (Lindsay 1979) and stimulate neurite outgrowth from dorsal root ganglion cells (Assouline et al. 1987). Additionally, cultures of embryonic Schwann cells express the receptor for NGF (NGFr) (Hosang and Shooter 1985; Rohrer 1985; Rohrer and Sommer 1983; Zimmermann and Sutter 1983). Thus, the evidence suggests a role for glial cells in the NGF response.

3 NGF Function on CNS Neurons In Vivo

A number of studies have shown that NGF occurs in and is produced by both the developing and adult CNS. In support of results obtained from in vitro developmental studies, NGF infusion to neonatal rats increased ChAT content in the basal forebrain and septum (Mobley et al. 1985, 1986; Johnson et al. 1987). NGF also appears to play a prominent role in the development of the septohippocampal cholinergic projection. The septohippocampal projection in the rat develops from embryonic day 20 to postnatal day 14 (Milner et al. 1983), providing the source of all extrinsic cholinergic innervation to the hippocampal formation. Total NGF content and NGF mRNA in the hippocampal formation rise just prior to developmental increases in ChAT activity during development, however, suggesting a role for NGF in the guidance and support of approaching septal cholinergic fibers (LARGE et al. 1986; WHITTEMORE et al. 1986). NGF receptor (NGFr) content (total protein) (ECKENSTEIN 1988; YAN and JOHNSON 1988), NGFr mRNA (BUCK et al. 1987), and total NGF content, but not NGF mRNA (WHITTEMORE et al. 1986), rise during embryonic and neonatal basal forebrain development, suggesting that these regions might be preparing to contact a source of, and retrogradely transport, NGF.

In vivo data also support in vitro data regarding the role of NGF in the development of cholinergic neurons of the nucleus basalis of Meynert (nucleus basalis). NGF administration to developing neurons in this region results in increases in ChAT activity (MOBLEY et al. 1986; JOHNSTON et al. 1987), although of a lesser degree than that seen in the septal region.

In the adult, mammalian CNS tissues have been shown to contain: NGF mRNA by northern (RNA) blotting and in situ hybridization (AYER-LELIEVRE et al. 1983; WHITTEMORE et al. 1986; LARGE et al. 1986; SHELTON and REICHARDT 1986; LARKFORS et al. 1987); NGF antigen by immunohistochemical and radioimmunoassays (AYER-LELIEVRE et al. 1983; GREENE and SHOOTER 1980); NGFr by autoradiography (RICHARDSON et al. 1986); and NGF by biological assays (COLLINS and DAWSON 1983; SCOTT et al. 1981). The highest NGF levels in CNS tissue appear within the target areas of the cholinoceptive basal forebrain systems (SHELTON and REICHARDT 1986), including the hippocampus and cortex, and NGF administered into rat brain raises ChAT levels in the hippocampus and septum (HEFTI et al. 1984; MOBLEY et al. 1985). Radiolabeled NGF injected into target regions is taken up and retrogradely transported by cholinergic neurons, including the septal/diagonal band neurons for the hippocampus and nucleus basalis neurons for the neocortex (SCHWAB et al. 1979; SEILER and SCHWAB 1984). Assays of NGF activity, mRNA synthesis, or NGFr positivity have also revealed NGF influence in adult olfactory bulb (FRIEDMAN et al. 1988; GOMEZ-PINILLA et al. 1988), posterior lobe of the pituitary gland (YAN et al. 1988), and neonatal and perhaps adult cerebellum (ECKENSTEIN 1988; TANIUCHI et al. 1986; PIORO et al. 1988).

Trophic factors, possibly including NGF also accumulate in fluids surrounding rat brain lesions (CRUTCHER 1987; COLLINS and CRUTCHER 1985; GAGE et al. 1984; NIETO-SAMPEDRO et al. 1983). Following lesions of the fimbria/fornix in the adult rat, total NGF content in the hippocampus rises by approximately 50%, although NGF mRNA does not rise (WHITTEMORE et al. 1986; KORSCHING et al. 1986). The latter finding suggests a buildup of NGF in the hippocampus from supplies of NGF that would normally have been retrogradely transported to cells of the medial septum. In the neonate, however, fimbria/fornix transections *are* associated with an elevation in hippocampal NGF mRNA (WHITTEMORE et al. 1986). The accumulation of NGF in the hippocampus after denervation may also result in the sprouting of another population of NGF-dependent processes into this region: axons of sympathetic neurons located in the superior cervical ganglion (CRUTCHER 1987). Cholinergic neurons of the adult striatum may also respond to NGF, primarily under conditions of trauma or stress (see below).

4 NGF Function on Damaged Adult Neurons: The Model

The cholinergic projection from the adult rat septum and diagonal band to the ipsilateral hippocampus has been a useful model for examining CNS plasticity (Fig. 1). Neurons of the medial septum and the vertical limb of the diagonal band project dorsally to the hippocampus mainly through the fimbria/fornix (GAGE et al. 1983; LEWIS et al. 1967; ARMSTRONG et al. 1987). About 50% of the septal/diagonal band neurons sending fibers through the fimbria/fornix are cholinergic (AMARAL and KURZ 1985; WAINER et al. 1985) and provide the hippocampus with about 90% of its total cholinergic innervation (STORM-MATHISEN 1974).

Complete transection of the fimbria/fornix pathway in adult rats results in the retrograde degeneration and death of many of the septal/diagonal band neurons that originally contributed axons through this pathway (ARMSTRONG et al. 1987; CUNNINGHAM 1982; GRADY et al. 1984; HEFTI 1986; WAINER et al. 1985). Markers of cell survival, including retrogradely transported fluorescent dyes and Nissl stains (GAGE et al. 1986; TUSZYNSKI et al. 1988), transmitter enzyme expression (AChE, ChAT) (GAGE et al. 1986; ARMSTRONG et al. 1987), and NGFr expression on cells of the medial septum (SPRINGER et al. 1987; MONTERO and HEFTI 1988; TUSZYNSKI et al. 1988), demonstrate a loss of 70%–90% of cells in this region. One recent study suggested that some of these cells may persist for extended periods in a dysfunctional state following lesions of the fimbria/fornix (HAGG et al. 1988), although several studies suggest that many of these cells actually die (MONTERO and HEFTI 1988; TUSZYNSKI et al. 1988; ARIMATSU et al. 1988; MV Sofroniew, personal communication).

One explanation for this axotomy-induced cell dysfunction or death is that the septal neurons become deprived of a critical supply of NTF possibly provided

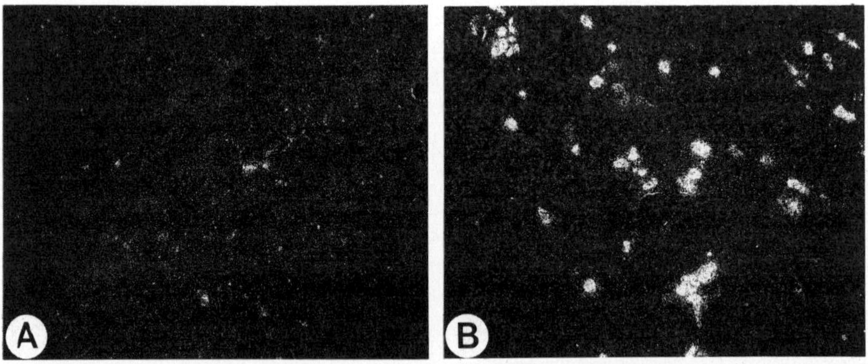

Fig. 1 A, B. NGF induction of NGF receptor mRNA within nucleus basalis neurons of the adult rat brain. Dark field photomicrographs show in situ hybridization of NGFr mRNA using a ^{35}S labeled RNA probe in a coronal tissue section. **A** Non-infused side and **B** NGF-infused side of the same tissue section. Magnification is the same in both photomicrographs (from HIGGINS et al. 1989) × 200

by the postsynaptic neurons or glial cells in the target areas of the hippocampus (COLLINS and CRUTCHER 1985; GAGE et al. 1986; GNAHN et al. 1983; NIETO-SAMPEDRO et al. 1983). That this hippocampal NTF might be NGF or NGF-like is supported by the previously listed studies from several laboratories reporting an NGF presence within the septohippocampal system.

Another model for the study of NGF function in the damaged adult brain is the nucleus basalis cholinergic projection. The nucleus basalis consists of a collection of magnocellular cholinergic neurons in the basal forebrain that provide diffuse, predominantly ipsilateral projections to most of the cerebral cortex (JOHNSTON et al. 1981, 1987). Following cortical destructive lesions, the cholinergic cell bodies of this region undergo atrophy as shown by ChAT immunoreactivity (SOFRONIEW et al. 1983, 1987) and decrements in ChAT activity (STEPHENS et al. 1985). Death of the cholinergic cell bodies in this region does not occur, however. That this system may also be influenced by NGF is suggested by the presence of high levels of NGF in the cortex (COLLINS and DAWSON 1983; WHITTEMORE et al. 1986; SHELTON and REICHARDT 1986; LARKFORS et al. 1987; SCOTT et al. 1984), the demonstration of retrograde transport of ^{125}I-NGF from cortex to nucleus basalis (SEILER and SCHWAB 1984), and the demonstration of NGFr on nucleus basalis neurons using immunocytochemistry and ^{125}I autoradiography (RICHARDSON et al. 1986; RAIVICH and KREUTZBERG 1987; SPRINGER et al. 1987; TANIUCHI and JOHNSON 1985; TANIUCHI et al. 1986; BATCHELOR et al. 1989; RIOPELLE et al. 1987).

5 NGF Function on Damaged Adult Neurons: The Effect

The above observations raise the question whether exogenous administration of NGF to neuronal populations in the adult showing NGF regulation might prevent lesion-induced neuronal degeneration and atrophy.

Within the septohippocampal circuit, for example, NGF administration to axotomized septal neurons might prevent retrograde cell degeneration and death. This rescue by NGF might then allow the axotomized neurons to regenerate their cut axons or extend new axons back to the hippocampal formation. Recently, several groups (HEFTI 1986; KROMER 1987; WILLIAMS et al. 1986; GAGE et al. 1988a) have independently reported that intraventricular administration of purified NGF into adult rats from the time of fimbria/fornix transection onward prevents the loss of most of the axotomized cholinergic septal/diagonal band neurons. Complete unilateral fimbria/fornix lesions usually result in a loss of 65%–90% of cholinergic cell bodies (GAGE et al. 1984; ARMSTRONG et al. 1987) compared to the contralateral, unlesioned side, but NGF infusion can rescue 90%–100% of the cell population (HEFTI 1986; KROMER 1987; WILLIAMS et al. 1986; GAGE et al. 1988a). Although various lesion and NGF infusion paradigms have been used in these studies, the results are consistent and comparable. NGF has typically been infused continuously into the lateral ventricle, ipsilateral to the fimbria/fornix lesion, for at least a 2-week period through an indwelling cannula that is connected to an osmotic minipump placed subcutaneously in the animal's back (WILLIAMS et al. 1986). Another experimental paradigm using twice weekly NGF injection into the ventricle of a partially lesioned animal will also prevent retrograde neuronal degeneration in the medial septum (HEFTI 1986), and unilateral NGF infusion after bilateral fimbria/fornix lesions will spare cholinergic neurons on *both* sides of the medial septum (KROMER et al. 1981). Both the 7S and purified 2.5S form of NGF have shown efficacy in these studies. A dose-response curve in this model system has not been determined, but efficacious doses employed empirically in the above studies have ranged from a few ug/ml to greater than 100 ug/ml (HEFTI 1986; KROMER 1987; WILLIAMS et al. 1986; GAGE et al. 1988a).

Recently, *transient* infusions of NGF have been shown to be insufficient to *permanently* rescue septal cholinergic neurons (MONTERO and HEFTI 1988). Thus work is currently directed toward inducing the regeneration of lesioned axons from rescued septal cells back to their hippocampal target in an attempt to restore an endogenous, continuous supply of NGF. NGF infusions could then be discontinued when axons reconnected with their target, thus permanently rescuing the projecting septal neurons (see below). The ability of NGF to induce axonal regeneration in the adult CNS in vivo has already been evidenced by the presence of robust sprouting of cholinergic fibers into the dorsolateral quadrant of the septum 2 weeks following fimbria/fornix lesions (GAGE et al. 1986, 1988a).

Intraventricular injections of NGF after partial bilateral fimbria/fornix lesions can ameliorate behavioral deficits. Following bilateral fimbria/fornix lesions, rats are impaired in tests of learning and mamory (OLTON et al. 1978). After chronic NGF treatment, lesioned rats are able to learn a radial arm maze as rapidly as control nonlesioned rats following an initial, immediate postoperative impairment (WILL and HEFTI 1985). In the same study an increase in ChAT activity is observed in the septum and occurs at the same time as the behavioral recovery. Also, the behavioral changes are not permanent; the behavioral deficits reappear after cessation of NGF treatment.

NGF is also capable of ameliorating some lesion-induced effects in the nucleus basalis model. Lesion-induced, retrograde atrophic changes in the nucleus basalis can be prevented by infusion of NGF into the ventricular system, resulting in restoration of nucleus basalis ChAT activity and augmentation of remaining ChAT activity in areas of cortex spared by the lesion (HAROUTUNIAN et al. 1986). Infusion of the monosialoganglioside GM1 will also restore nucleus basalis and cortical ChAT levels, but the combined infusion of both substances will augment the response further (CUELLO et al. 1989; GAROFALO et al. 1988). These authors suggested that GM1 may act as a modulator of NGF function.

NGF effects on cells other than cholinergic neurons in the adult CNS have received less attention. GABAergic cells in addition to cholinergic cells of the medial septum degenerate after fimbria/fornix lesions (PETERSON et al. 1987), and one recent study reports a failure of NGF to save these GABAergic neurons

(MONTERO and HEFTI 1988). The identification of NGF-responsive cell populations in the CNS depends in part upon the visualization of NGFr with immunocytochemical methods. However, since immunocytochemistry currently employs antibodies that appear to identify only the low affinity NGFr (TANIUCHI and JOHNSON 1985), future studies may require methods that detect the active, high affinity receptor as well (e.g., autoradiography).

6 NGF Receptor Expression is Regulated by NGF

The ability of neurons to respond to NGF seems to depend on the presence of cell surface receptors, which in the PNS mediate the binding, internalization, and transport of NGF from the terminals to the parent cell bodies (see THOENEN and BARDE 1980 for review). Such NGFr have also been demonstrated on NGF-responsive cholinergic neurons in the CNS both during development and in the adult animal (TANIUCHI et al. 1986; RICHARDSON et al. 1986).

The cholinergic neurons of the striatum appear to represent a special case. In contrast to the neurons in the septal diagonal band area and the nucleus basalis, these neurons possess very low or undemonstrable levels of NGFr in the adult, and their responsiveness to NGF has been reported to decline dramatically during postnatal development (MARTINEZ et al. 1985; MOBLEY et al. 1985; JOHNSTON et al. 1987). Nevertheless, our previous findings showed that chronic infusion of NGF into the lateral ventricle of adult animals following fimbria/fornix lesion not only spared the medial septal neurons from degeneration, but also resulted in hypertrophy of the cholinergic neurons of the ipsilateral striatum (GAGE et al. 1988b). Similarly, chronic infusions of NGF into the lateral ventricle of aged rats ameliorated the age-related atrophy of the cholinergic neurons of the striatum as well as of the basal forebrain (FISCHER et al. 1987). The in vivo effects of NGF on the striatal cholinergic neurons are in apparent contradiction with the lack of demonstrable NGFr immunoreactivity in these neurons normally. However, TANIUCHI et al. (1986, 1988) have recently reported that peripheral nerve damage will induce the expression of NGFr on Schwann cells within the denervated distal portion of the nerve, raising the possibility that the ability of striatal neurons to respond to NGF depends on the upregulation of NGFr and that this upregulation is induced by the tissue damage. In a recent experiment we tested this hypothesis. Chronic NGF infusion into the adult neostriatum resulted in re-expression of NGFr such that many cholinergic interneurons became immunoreactive for NGFr. This effect was seen also after striatal damage induced by infusion of vehicle alone, whereas infusion of anti-NGF serum partially inhibited the receptor's re-expression. Infusion of NGF, but not vehicle alone, dramatically increased the size and ChAT immunoreactivity of these same cholinergic neurons (BATCHELOR et al. 1989).

These findings indicate that the central cholinergic neurons which lose their NGFr during postnatal development will resume their NGF responsiveness

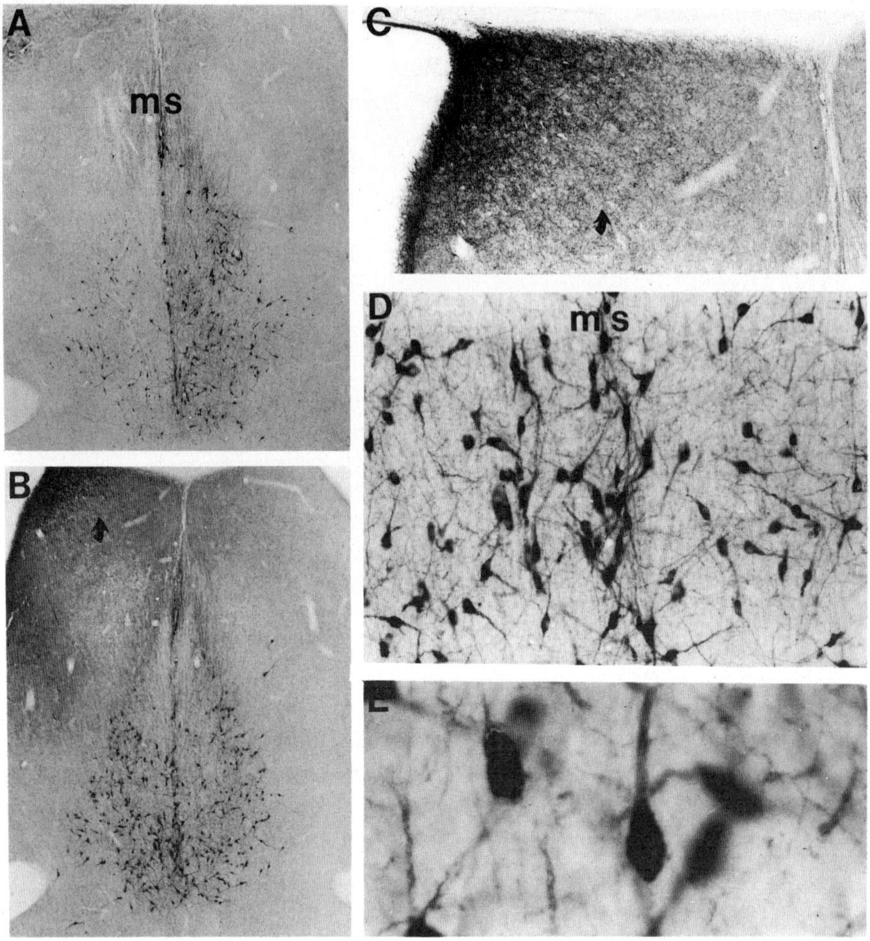

Fig. 2 A–E. Medial septal region after fimbria/fornix transection. **A** Appearance of medial septal cholinergic cells after unilateral fimbria/fornix transection. Lesion on *left* side, intact septum on *right*. Note the loss of total cell number and intensity of ChAT staining on the side of the lesion (ChAT immunocytochemistry) × 40. **B** Appearance of medial septal region after NGF infusion. Lesion on *left* side. Extensive cholinergic cell savings are evident. *Arrow* in upper left corner indicates sprouting of axons into dorsolateral quadrant of septum (ChAT) × 40. **C** Higher power view of sprouting into dorsolateral quadrant × 100. **D** Normal morphology of ChAT positive fibers in the medial septal region after unilateral fimbria/fornix lesions and NGF infusion. Lesion on *left* side, intact septum on *right* (ChAT) × 100. **E** High power view showing normal morphology of NGF-treated neurons (ChAT) × 1000

when the tissue is damaged. Such a damage-induced mechanism may act to enhance both the action of trophic factors, including NGF, released at the site of injury and the responsiveness of damaged CNS neurons to exogenously administered trophic factors.

Although damage produces an emergence of NGFr protein immunoreactivity in adult striatal cholinergic neurons, in situ hybridization of NGFr mRNA shows that vehicle infusion by itself will not increase the number of NGFr mRNA positive neurons in the striatum as compared to treated animals (GAGE et al. 1989a). Thus, only NGF treatment, but not vehicle, induces NGFr gene expression within striatal neurons, suggesting the possibility that lesion-induced damage increases NGFr levels through non-transcriptional mechanisms, such as the "unmasking" of existing stores of NGFr protein. Additionally, these findings are in agreement with our recent in situ hybridization studies which showed that chronic NGF administration, but not vehicle infusions, induce NGFr gene expression within those basal forebrain cholinergic neuronal populations that normally express NGFr protein in adult rat brain (HIGGINS et al. 1989). NGF not only produced neuronal hypertrophy, as we demonstrated in earlier studies (FISCHER et al. 1987), but also increased NGFr mRNA abundance per cell and induced the expression of NGFr message within basal forebrain neurons that had previously not contained detectable levels of NGFr mRNA (Fig. 2). These findings provide strong evidence that NGF can induce expression of its own receptor within the CNS.

In vitro evidence also supports the concept that NGF can induce the expression of its own receptor. When NGF is applied to cultures of septal neurons, striatal neurons, or PC12 cells (BERND and GREENE 1984; HEFTI and GAHWILER 1988), the number of NGFr on the cells increases. In addition, the exposure of cultured chick sensory neurons to long-term NGF treatment will prevent the normal disappearance of receptors on these cells (ROHRER and BARDE 1982).

7 NGF Function in the Aged Brain

In addition to the lesion-induced degeneration described above, several laboratories have demonstrated that, in aged animals and humans, the cholinergic neurons of the basal forebrain are compromised (BARTUS et al. 1982; WHITEHOUSE et al. 1982). This compromise is reflected in cell shrinkage and, in some cases, loss of cholinergically marked neurons in the basal forebrain region, which seems to correlate with a decrease in cognitive ability in animals (FISCHER et al. 1987) and humans (COYLE et al. 1983). RNA blotting and protein data suggest a reduction of NGF levels in the hippocampus of aged rats compared to the levels in adult rats (LARKFORS et al. 1987).

We have attempted to ameliorate the functional deficits and reverse the morphological changes observed in aged animals by implanting fetal basal forebrain cells or by infusing NGF chronically into the aged brain. We initially reported and subsequently confirmed and extended the findings that transplantation of fetal cholinergic neurons to the hippocampal formation of aged rats prescreened for cognitive impairments could result in substantial improvement in the previously impaired animal's behavior and that this was in part mediated via the cholinergic system (GAGE et al. 1983; GAGE and BJORKLUND 1986b).

Table 1. AChE positive cell body size and spatial memory retention (modified from FISCHER et al. 1987)

		MS	VDB	Striatum	Nucleus basalis	Retention
Aged impaired, NGF-treated	L	129.8 ± 9.5	136.6 ± 7.9	122.0 ± 10.7	155.6 ± 6.2	6.6 ± 1.0**
	R	136.8 ± 8.8	143.4 ± 5.6	156.2 ± 11.7*	184.8 ± 10.0*	
Aged impaired, vehicle-treated	L	119.9 ± 5.3	146.4 ± 6.3	140.1 ± 6.5	165.4 ± 3.9	24.4 ± 3.5
	R	132.0 ± 7.4	133.9 ± 10.0	143.4 ± 4.8	162.1 ± 3.8	
Young	L	170.2 ± 7.4	226.6 ± 11.2	236.8 ± 6.6	268.8 ± 23.4	
	R	172.7 ± 2.6	229.5 ± 16.7	233.3 ± 3.8	270.1 ± 18.9	

Values are means ± SEM of AChE positive cell body size given as cross-sectional area in μm^2. The mean escape latency (± SEM) in the memory retention test on day 1 of the second test week is also given for the two subgroups of impaired rats.
MS, medial septum; VDB, vertical limb of diagonal band; L, non-infused side; R, infused side
* $P < 0.01$ compared to the non-infused side; Student's related t test
** $P < 0.01$ compared to the vehicle injected controls; Student's unrelated t test

Additionally, based on our initial observations of the trophic effect of NGF on cholinergic neurons in the retrograde degeneration model described above, we infused NGF into the lateral ventricles of cognitively impaired aged rats and found that all aged rats with intact NGF pumps showed an improvement in retention of a complex spatial learning task relative to matched, noninfused, cognitively impaired rats (FISCHER et al. 1987). In the same study, we observed a significant increase in the size of the cholinergic neurons in the basal forebrain region on the side of the brain into which the NGF was intraventricularly infused (Table 1).

8 NGF Functions as a Tropic Factor

In addition to its function of maintaining the normal integrity of central cholinergic basal forebrain neurons and of peripheral sympathetic and neural crest-derived sensory neurons (*neurotrophic*), NGF has been postulated to have a role in the axonal growth of these neurons after damage to the nervous system (*neurotropic*). Two forms of axonal growth are commonly observed in the mature hippocampus: *regeneration*, or the regrowth of axons previously damaged, and *collateral sprouting*, or the new growth of remaining (intact) axons (GAGE and BJORKLUND 1986b). Collateral sprouting occurs within the hippocampus from two distinct populations of neurons following fimbria/fornix transection. Superior cervical ganglion (SCG)-derived sympathetic axons, which normally surround the hippocampal vasculature, undergo a robust sprouting response into the dentate gyrus and CA3 pyramidal cell region of the hippocampal parenchyma (LOY and MOORE 1977; STENEVI and BJORKLUND 1978). In addition, the magnocellular midline cholinergic neurons on the dorsal hippocampal formation respond by sprouting into the dorsal subiculum and CA1 pyramidal cell layers (BLAKER et al. 1988). To test the postulate that NGF has tropic effects on cholinergic axons in the hippocampus and that the lesion-induced increase in NGF in the hippocampal formation may serve as a chemoattractant of NGF-responsive axons towards the source of its production (CRUTCHER 1987), we examined the fimbria/fornix lesioned hippocampal formation to determine whether the sprouting neurons stain positively for NGFr. We found that the two populations of neurons which undergo collateral sprouting, namely, the midline magnocellular cholinergic neurons of the dorsal hippocampus and the sympathetic neurons of the SCG, stain strongly for NGFr with a monoclonal antibody (TANIUCHI et al. 1986, 1988). In contrast, the small intrinsic cholinergic neurons of the hippocampus exhibited neither sprouting response nor staining for NGFr (BATCHELOR et al. 1989). In view of these results we suggest that the differing sprouting responses demonstrated by these three neuronal populations may be due to their responsiveness to NGF, as indicated by the presence or absence of NGFr.

9 NGF Induces Regeneration in the CNS

Attempts to restore this severed fimbria/fornix pathway have been made by grafting fetal tissue or other substances that may act as bridges (TUSZYNSKI et al. 1990; GAGE et al. 1988b; KROMER et al. 1981) between the disconnected septohippocampal pathway. Increases in ChAT activity and AChE fiber innervation in the host hippocampus have been consistently reported in these studies, but the extent of reinnervation is small.

A possible explanation for the limited restoration of the cholinergic circuitry may be that the majority of the cholinergic neurons in the medial septum and diagonal band of Broca degenerate, become dysfunctional, and die within a month following the transection (DAITZ and POWELL 1954; GAGE et al. 1986; ARMSTRONG et al. 1987). This observation, and the evidence of a link between NGF and cholinergic neurons (HEUMANN et al. 1985; HONEGGER and LENOIR 1982; SHELTON and REICHARDT 1986), have prompted several groups to test and subsequently demonstrate the dependence of adult denervated cholinergic neurons on exogenous NGF for survival in the absence of endogenous NGF previously transported from the hippocampus (GAGE et al. 1988a; HEFTI 1986; KROMER 1987; WILLIAMS et al. 1986).

These results led to the prediction that the exogenous delivery of NGF could not only promote the survival of septal neurons, but also would then promote cholinergic axon extension across a bridge of hippocampal fetal tissue placed in the fimbria/fornix cavity. Thus, in a recent study we combined the exogenous infusion of NGF into the lateral ventricle adjacent to the denervated septum with the simultaneous grafting of fetal hippocampal tissue to the fimbria/fornix cavity as a set of procedures that may more fully and functionally restore the severed septohippocampal circuitry (TUSZYNSKI et al. 1990). Embedded in the design of this study were two, additional, related questions: (1) Does the transient 2 week NGF infusion period, which has been shown to result in significant cholinergic cell rescue, have an enduring effect on the medial septal cells 6–8 months following termination of the NGF infusion? (2) Will fetal hippocampal grafts alone, in the absence of exogenous NGF infusion, support the survival of the axotomized cholinergic neurons of the medial septum?

A combination of intracerebral grafting and intraventricular infusion of NGF was used to attempt reconstruction of the cholinergic component of the septohippocampal pathway following fimbria/fornix lesions. Four groups were used: lesion only (FF); lesion and fetal hippocampal graft (FF-HPC); lesion and NGF (FF-NGF); and lesion, graft, and NGF (FF-HPC-NGF). ChAT immunoreactivity (ChAT-IR), AChE fiber staining, and behavior-dependent theta electrical activity were used to assess the extent of pathway reconstruction. The NGF infusion only lasted the first 2 weeks following the FF lesion, while measurement of theta activity and histological analysis were conducted 6–8 months after lesioning. Only the FF-HPC-NGF group had long-term rescue of ChAT-IR cells as compared to the FF and FF-HPC group. In addition, the FF-HPC-NGF group had more extensive reinnervation of the hippocampus than any other group. Further, the FF-HPC-NGF had the most complete evidence of behavior-dependent theta activity restoration. These results demonstrate clearly that a combination of short-term intraventricular NGF infusion and fetal hippocampal grafts can result in a more complete reconstruction of the damaged septohippocampal circuit.

10 A Role for Glia in Mediating the Tropic Effects of NGF

Damage to the fimbria/fornix and separately to the perforant path leads to distinct and dramatic time dependent increases in glial fibrillary acidic protein immunoreactivity (GFAR-IR) in specific areas of the hippocampal formation (GAGE et al. 1988c). Specifically, FF lesions resulted in an increase in GFAP-IR in the pyramidal and stratum oriens layers of the CA3 region as well as in the inner molecular layer of the dentate gyrus. In addition, in the septum ipsilateral to the lesion, there was a rapid and robust increase in GFAP-IR in the dorsal lateral quadrant of the septum, but not in the medial septal region. Only after 30 days did GFAR-IR reach the medial septum. Following perforant path lesions, there was a selective increase in GFAP-IR in the outer molecular layer of the dentate gyrus. Most of these changes were transient and disappeared by 30 days post-lesion. We speculated that the increase in GFAP-IR, reflecting activation of astrocytic cells in these target areas, is a necessary requirement for sprouting responses (GAGE et al. 1988c).

Considerable in vitro and in vivo evidence supports the presumption that astrocytes make and secrete NTF which can subsequently support the survival and/or axonal outgrowth of a variety of central and peripheral neurons (Banker 1980; Hatten and Liem 1981; Liesi et al. 1984; Lindsay 1979; Tarris et al. 1986). At present, when a new putative factor is tested in vitro for its neurotrophic activity, rigorous controls must be used to establish that the neuronal population is not contaminated with glia and that the presumed NTF is not acting through the glial cell population in vitro. To date, the evidence supports the notion that only reactive and/or proliferating astrocytes secrete trophic and tropic substances; thus, it is essential to understand the signals for this activation in vivo. Microglial proliferation often precedes astrocytic proliferation (Gall et al. 1979; Vijayan and Cotman 1983). Recently, Giulian et al. (1986) and Hetier et al. (1988) showed that activated microglia produce interleukin-1 (IL-1) in vitro and that IL-1 stimulates astrocyte proliferation in vitro and in vivo (Giulian et al. 1988). Very recently it has been shown that IL-1 regulates the synthesis of NGF in non-neuronal cells of the damaged rat sciatic nerve (Lindholm et al. 1987) and that IL-1 is most likely secreted from activated macrophages in the vicinity of the damaged nerve (Heumann et al. 1987b).

Work in the CNS has demonstrated time dependent changes in microglial and astrocytic populations in the outer molecular layer of the dentate gyrus denervated by perforant path transection (Fagan et al. 1988). Reactive microglia stained with the monoclonal antibody OX-42 appear in this area within 24 h after transection. The appearance of cells immunostained for IL-1 in this area parallels that of microglia, suggesting that the reactive microglia themselves produce IL-1 in vivo in response to deafferentation. The astrocyte response, as indexed by GFAP-IR, is not observed until a later time point coincident with the disappearance of IL-1 staining. These observations provide evidence that reactive microglia produce IL-1 in vivo in response to injury and that IL-1 may be the signal responsible for the documented astrocyte response in the outer molecular layer of the dentate gyrus observed after damage to the entorhinal input to the hippocampus.

11 Working Hypothesis of NGF Effects

The results of data summarized in the previous sections suggest: (1) a role for IL-1 in the proliferation of astrocytes, (2) that activated microglia and macrophages can secrete IL-1, and (3) that IL-1 can activate NGF synthesis in non-neuronal cells. We have made the following suggestion for the outline of events that lead to the NGF-sensitive sprouting responses in the hippocampus and septum following fimbria/fornix and perforant pathway lesions.

Perforant pathway damage induces terminal degeneration of the axons of cells which were transected in the entorhinal cortex. This terminal degeneration in the outer molecular layer of the dentate gyrus activates microglia to phagocytize in the restricted zone of terminal degeneration. These activated microglia release IL-1 into the surrounding environment, which induces the proliferation of astrocytes. The activated astrocytes, in turn, secrete NGF which results in the attraction of cholinergic fibers that express NGFr on their membrane surfaces.

Similarly, following fimbria/fornix lesions, cholinergic terminals are disconnected from their cell bodies in the septum and, once again, there is terminal degeneration, this time in the areas of heaviest cholinergic innervation in the CA3 region and dentate gyrus. This degeneration leads to a microglial proliferation, IL-1 secretion, and an astrocytic mitogenic reaction in CA3 and inner molecular layer of the dentate gyrus. The activated astrocytes secrete NGF and promote the ingrowth of NGFr-bearing sympathetic fibers of the SCG.

Concurrently in the septum, degeneration of terminals from the hippocampus to the dorsolateral quadrant results in microglial reactivity and subsequent IL-1 secretion, proliferation of glia, and increased NGF concentration. This, in turn, induces the growth of cholinergic fibers from the medial septum, which are undergoing retrograde degeneration, into the dorsolateral quadrant. This NGF source in the dorsolateral quadrant is not sufficient to support all the medial cholinergic cells, but some of the medial septal cells are always spared even with a complete bilateral fimbria/fornix lesion. Meanwhile, the absence of reactive astrocytes in the medial septum soon after the fimbria/fornix lesion is the paramount reason for the death of these cells, because, as in development, these cells survive only in the presence of adequate glia-derived NGF.

This working hypothesis generates several specific testable predictions, which when examined should reveal more about the mechanisms underlying survival, growth, and function of normal and damaged cholinergic neurons in the CNS.

12 Conclusions

Based on this review, the following points summarize our knowledge of NGF function in the CNS:

1. NGF is synthesized in target areas of the projecting cholinergic neurons of the forebrain.
2. NGFr is synthesized in basal forebrain cholinergic projecting neurons and anterogradely transported to its terminals in target areas which produce NGF.
3. Survival of the projecting basal forebrain neurons is dependent on the retrograde transport of NGF to basal forebrain neurons.
4. Disconnection of the projecting neurons from the targets results in retrograde degeneration which can be reversed by exogenous administration of NGF.
5. NGF exogenous infusion can also promote axonal regrowth, provided a substrate for axonal growth is available.

6. NGF appears to have a positive effect on its own receptors, such that increased levels of NGF result in increased expression of NGFr and increased responsiveness to NGF.
7. In the aged rat with cognitive impairments in behavior which are correlated with cholinergic cell size and number, chronic NGF infusion can result in improved behavioral performance and partial restoration of cholinergic cell size.
8. While NGF appears to be synthesized in neurons of the intact target areas, damage-induced activated astroglia may also have the ability to synthesize and secrete NGF.
9. IL-1 may regulate NGF expression and activate astrocytes; this regulation of NGF may contribute to sprouting responses of cholinergic neurons in the CNS.

While a complete understanding of NGF regulation and function is not available, both a trophic and tropic role for NGF exists in the CNS. The specificity of NGF for cholinergic neurons, while not perhaps exact, may be unique in its specificity. To date, the other candidate neurotrophic molecules have a broader range of function. Thus while NGF may be the first and best characterized NTF, it may not be the prototype for other trophic factors in the CNS.

Acknowledgement. We thank Sheryl Christenson for typing the manuscript.

References

Amaral DG, Kurz J (1985) An analysis of the origins of the cholinergic and noncholinergic septal projections to the hippocampal formation of the rat. J Comp Neurol 240: 37–59

Appel SH (1981) A unifying hypothesis for the cause of amyotrophic lateral sclerosis, parkinsonism, and Alzheimer disease. Ann Neurol 10: 499–505

Arimatsu Y, Miyamoto M, Tsukui H, Hatanaka H (1988) Nerve growth factor enhances survival of identified projection neurons in the rat septal and diagonal band regions in vitro. Soc Neurosci Abstr 14: 1114

Armstrong DM, Terry RD, Deteresa RM, Bruce G, Hersh LB, Gage FH (1987) Response of septal cholinergic neurons to axotomy. J Comp Neurol 264: 421–436

Assouline JG, Bosch P, Lim R, Kim IS, Jensen R, Pantazis NJ (1987) Rat astrocytes and Schwann cells in culture synthesize nerve growth factor-like neurite-promoting factors. Dev Brain Res 31: 103–118

Ayer-LeLievre CS, Ebendal T, Olsen L, Seiger A (1983) Localization of NGF-like immunoreactivity in rat neurons tissue. Med Biol 61: 296–304

Banker GA (1980) Tropic interactions between astroglial cells and hippocampal neurons in cultures. Science 209: 809–811

Barde YA, Edgar D, Thoenen H (1983) New neurotrophic factors. Annu Rev Physiol 45: 601–612

Bartus R, Dean RL, Beer C, Lippa AS (1982) The cholinergic hypothesis of geriatric memory dysfunction. Science 217: 408–417

Batchelor PE, Armstrong DM, Blaker SM, Gage FH (1989) Nerve growth factor receptor and choline acetyltransferase colocalization in neurons within the rat forebrain: response to fimbria-fornix transection. J Comp Neurol 284: 187–204

Berg DK (1984) New neuronal growth factors. Annu Rev Neurosci 7: 149–170
Bernd P, Greene LA (1984) Association of ^{125}I-nerve growth factor with PC12 pheochromocytoma cells. Evidence for internalization via high-affinity receptors only and for long-term regulation by nerve growth factor. J Biol Chem 259: 15509–15516
Blaker SN, Armstrong DM, Gage FH (1988) Cholinergic neurons within the rat hippocampus: response to fimbria-fornix transection. J Comp Neurol 272: 127–138
Buck CR, Martinex HJ, Black IB, Chao MV (1987) Developmentally regulated expression of the nerve growth factor receptor gene in the periphery and brain. Proc Natl Acad Sci USA 84: 3060–3063
Buzsaki G, Bickford RG, Varon S, Armstrong DM, Gage FH (1987) Reconstruction of the damaged septohippocampal circuitry by a combination of fetal grafts and transient NGF infusion. Soc Neurosci Abstr 13: 568
Burnstock G (1974) Degeneration and regeneration of growth of autonomic nerves in relation to smooth muscle in joint tissue cultures and anterior eye chamber transplants. In: Fuxe K, Olson L, Zollerman Y (eds) Dynamics of degeneration and growth in neurons. Pergamon, Oxford, pp 509–520
Butcher LL (1983) Acetylcholinesterase histochemistry. Elsevier, Amsterdam, pp 1–49 (Handbook of chemical neuroanatomy, vol 1)
Campenot RB (1977) Local control of neurite development by nerve growth factor. Proc Natl Acad Sci USA 74: 4516–4519
Chamley JH, Goller I, Burnstock G (1973) Selective growth of sympathetic nerve fibres to explants of normally densely innervated autonomic effector organs in tissue culture. Dev Biol 31: 362–379
Charlwood KA, Lamont DM, Banks BEC (1972) Apparent orienting effects produced by nerve growth factor. In: Zamis E, Kinight J (eds) Nerve growth factor and its antiserum. Athlone, London, pp 102–107
Chun LLY, Patterson PH (1977a) Role of nerve growth factor in the development of rat sympathetic neurons in vitro. I. Survival, growth, and differentiation of catecholamine production. J Cell Biol 75: 694–704
Chun LLY, Patterson PH (1977b) Role of nerve growth factor in the development of rat sympathetic neurons in vitro. II. Developmental study. J Cell Biol 75: 705–711
Chun LLY, Patterson PH (1977c) Role of nerve growth factor in the development of rat sympathetic neurons in vitro. III. Effect on acetylcholine production. J Cell Biol 75: 712–718
Collins F, Crutcher KA (1985) Neurotrophic activity in the adult rat hippocampal formation: regional distribution and increase after septal lesion. J Neurosci 5: 2809–2814
Collins F, Dawson A (1983) An effect of nerve growth factor on parasympathetic neurite outgrowth. Proc Natl Acad Sci USA 80: 2091–2094
Cowan WM, Fawcett JW, O'Leary DD, Stanfield BB (1984) Regressive events in neurogenesis. Science 225: 1258–1265
Coyle JT, Price PH, Delong MR (1983) Alzheimer's disease: a disorder of cortical cholinergic innervation. Science 219: 1184–1189
Crain SM (1975) Physiology of CNS tissue in culture. In: Berl S, Clarke DD, Schneider DD (eds) Metabolic compartmentation and neurotransmission. Relation to brain structure and function. Plenum, New York, pp 273–303
Crain SM, Peterson ER (1974) Enhanced afferent synaptic functions in fetal mouse spinal cord-sensory ganglion explants following NGF-induced ganglion hypertrophy. Brain Res 79: 145–152
Crutcher KA (1987) Sympathetic sprouting in the central nervous system: a model for studies of axonal growth in the mature mammalian brain. Brain Res Rev 12: 203–233
Crutcher KA, Collins F (1982) In vitro evidence for two distinct hippocampal growth factors: basis of neuronal plasticity? Science 217: 67–68
Cuello AC, Maysinger D, Garofalo L, Tagari P, Stephens PH, Pioro E, Piotte M (1989) Influence of gangliosides and nerve growth factor on the plasticity of forebrain cholinergic neurons. In: Fuxe K, Agnati LF (eds) Receptor-receptor interactions. (Proceedings of the Wenner-Gren symposium.) Macmillan, Basingstoke, pp 62–77
Cunningham TJ (1982) Naturally occurring neuron death and its regulation by developing neural pathways. Int Rev Cytol 74: 163–186
Daitz HM, Powell TPS (1954) Studies on the connexions of the fornix system. J Neurol Neurosurg Psychiatry 7: 75–82
Dreyfus CF, Peterson ER, Crain SM (1980) Failure of nerve growth factor to affect fetal mouse brain stem catecholaminergic neurons in culture. Brain Res 194: 540–547

Ebendal T, Jacobson CO (1977) Tissue explants affecting extension and orientation of axons in cultured chick embryo ganglia. Exp Cell Res 121: 353–361
Eckenstein F (1988) Transient expression of NGF-receptor-like immunoreactivity in postnatal rat brain and spinal cord. Brain Res 446: 149–154
Fagan AM, Robertson R, Gage FH (1988) Degeneration and regeneration in the outer molecular layer of the dentate gyrus after perforant path lesions: role of astrocytes, microglia and interleukin-1. Soc Neurosci Abstr 14: 116
Fawcett JW, O'Leary DDM, Cowan WM (1984) Activity and the control of ganglion cell death in the rat retina. Proc Natl Acad Sci USA 81: 5589–5593
Fischer W, Wictorin K, Bjorklund A, Williams LR, Varon S, Gage FH (1987) Amelioration of cholinergic neuron atrophy and spatial memory impairment in aged rats by nerve growth factor. Nature 329: 65–68
Fischer W, Gage FH, Bjorklund A (1989) Degenerative changes in forebrain cholinergic nuclei correlate with cognitive impairments in aged rats. Eur J Neurosci 1: 34–45
Fonnum F (1969) Radiochemical microassays for the determination of choline acetyltransferase and acetylcholinesterase activities. J Biochem 115: 465–472
Fonnum F (1984) Topographical and subcellular localization of choline acetyltransferase in the rat hippocampal region. J Neurochem 24: 407–409
Friedman PL, Larhammar D, Holets VR, Gonzalez-Cavajal M, Yu ZY, Persson H, Whittemore SR (1988) Rat β-NGF sequence and sites of synthesis in adult CNS. Soc Neurosci Abstr 14: 827
Gage H, Bjorklund A (1986a) Neural grafting in the aged rat brain. Ann Res Physiol 48: 447–459
Gage FH, Bjorklund A (1986b) Cholinergic septal grafts into the hippocampal formation improve spatial learning and memory in aged rats by an atropine sensitive mechanism. J Neurosci 6: 2837–2847
Gage FH, Bjorklund A (1986c) Compensatory collateral sprouting of aminergic systems in the hippocampal formation following partial deafferentation. In: Isaacson RL, Pribram KH (eds) The hippocampus, vol 3. Plenum, New York, pp 33–63
Gage FH, Bjorklund A (1986d) Enhanced graft survival in the hippocampus following selective denervation. Neuroscience 17: 89–98
Gage FH, Bjorklund A, Stenevi U, Dunnett SB (1983) Functional correlates of compensatory collateral sprouting by aminergic and cholinergic afferents in the hippocampal formation. Brain Res 268: 39–47
Gage FH, Bjorklund A, Stenevi U (1984) Denervation releases a neuronal survival factor in adult rat hippocampus. Nature 308(12): 637–639
Gage FH, Wictorin K, Ficher W, Williams LR, Varon S, Bjorklund A (1986) Life and death of cholinergic neurons in the septal and diagonal band region following complete fimbria-fornix transection. Neuroscience 19: 241–255
Gage FH, Armstrong DM, Williams LR, Varon S (1988a) Morphologic response of axotomized septal neurons to nerve growth factor. J Comp Neurol 269: 147–155
Gage FH, Blaker SN, Davis GE, Engvall E, Varon S, Manthorpe M (1988b) Human amnion membrane matrix as a substratum for axonal regeneration in the central nervous system. Exp Brain Res 72: 371–380
Gage FH, Olejniczak P, Armstrong DM (1988c) Astrocytes are important for NGF-mediated hippocampal sprouting. Exp Neurol 102: 2–13
Gage FH, Batchelor P, Chen KS, Chin D, Higgins GA, Koh S, Deputy S, Rosenberg MB, Fischer W, Bjorklund A (1989) NGF receptor re-expression and NGF mediated cholinergic neuronal hypertrophy in the damaged adult neostriatum. Neuron 2: 1177–1187
Gahwiler BH, Enz A, Hefti F (1987) Nerve growth factor promotes development of the art septo-hippocampal cholinergic projection in vitro. Neurosci Lett 75: 6–10
Gall C, Rose G, Lynch G (1979) Proliferative and migratory activities of glial cells in the partially deafferented hippocampus. J Comp Neurol 183: 539–550
Garofalo L, Maysinger D, Cuello AC (1988) In vivo effects of GM1 and NGF, administered in combination, following CNS retrograde degeneration. Soc Neurosci Abstr 14: 826
Giulian D, Baker TJ, Shih LN, Lachman LB (1986) Interleukin-1 of the central nervous system is produced by ameboid microglia, J Exp Med 164: 594–604
Giulian D, Woodward J, Young DG, Krebs JF, Lachman LB (1988) Interleukin-1 injected into mammalian brain stimulates astrogliosis and neovascularization. J Neurosci 8: 2485–2490
Gnahn H, Hefti F, Heumann R, Schwab ME, Thoenen H (1983) NGF-mediated increase in choline acetyltransferase (ChAT) in the neonatal rat forbrain; evidence for physiological role of NGF in the brain? Dev Brain Res 9: 45–52

Gomez-Pinilla F, Guthrie K, Leon M, Nieto-Sampedro M (1988) NGF receptor increase in the olfactory bulb after early odor deprivation in rats. Soc Neurosci Abstr 14: 685

Grady S, Reeves T, Steward O (1984) Time course of retrograde degeneration of the cells of origin of the septohippocampal pathway after fimbria-fornix transections. Soc Neurosci Abstr 10: 463

Greene LA (1977a) Quantitative in vitro studies on the nerve growth factor requirement of neurons. I. Sympathetic neurons. Dev Biol 58: 96–105

Greene LA (1977b) Quantitative in vitro studies on the nerve growth factor requirement of neurons. II. Sensory neurons. Dev Biol 58: 106–113

Greene LA, Shooter EM (1980) The nerve growth factor: biochemistry, synthesis, and mechanism of action. Annu Rev Neurosci 3: 353–402

Gundersen RW (1985) Sensory neurite growth cone guidance by substrate adsorbed nerve growth factor. J Neurosci Res 13: 199–212

Gundersen RW, Barrett JN (1979) Neuronal chemotaxis: chick dorsal root axons turn toward high concentrations of nerve growth factor. Science 206: 1079–1080

Gundersen RW, Barrett JN (1980) Characterization of the turning response of dorsal root neurites toward nerve growth factor. J Cell Biol 87: 546–554

Hagg T, Manthorpe M, Vahlsing HL, Varon S (1988) Delayed treatment with nerve growth factor reverses the apparent loss of cholinergic neurons after brain damage. Exp Neurol 101: 303–312

Hamburger V, Oppenheim RW (1982) Naturally occurring neuronal death in vertebrates. Neurosci Comment 1: 55–68

Haroutunian V, Kanof PD, Davis KL (1986) Partial reversal of lesion-induced deficits in cortical cholinergic markers by nerve growth factor. Brain Res 386: 397–399

Hartikka J, Hefti F (1988) Development of septal cholinergic neurons in culture: plating density and glial cells modulate effects of NGF on survival, fiber growth, and expression of transmitter-specific enzymes. J Neurosci 8(8): 2967–2985

Hatanaka H, Tsukui H (1986) Differential effects of nerve growth factor and glioma-conditioned medium on neurons cultured from various regions of fetal rat central nervous system. Dev Brain Res 30: 47–56

Hatten ME, Liem RHK (1981) Astroglial cells provide a template for the positioning of developing cerebellar neurons in vitro. J Cell Biol 90: 622–630

Hefti F (1986) Nerve growth factor (NGF) promotes survival of septal cholinergic neurons after fimbrial transection. J Neurosci 6: 2155–2162

Hefti F, Dravid A, Hartikka JJ (1984) Chronic intraventricular injections of nerve growth factor elevate hippocampal choline acetyltransferase activity in adult rats with partial septohippocampal lesions. J Brain Res 293: 305–311

Hefti F, Hartikka JJ, Eckenstein F, Gnahn H, Heumann R, Schwab M (1985) Nerve growth factor increases choline acetyltransferase but not survival or fiber outgrowth of cultured fetal septal cholinergic neurons. Neurosci 14: 55–68

Hefti F, Gahwiler BH (1988) Cholinergic neurons of the rat forebrain in slice cultures: interactions with target tissue and effects of nerve growth factor. In: Gorio A, Perez-Pete R (eds): Neural

Hetier E, Ayala J, Denefle P, Bousseau A, Rouget P, Mallat M, Prochiantz A (1988) Brain macrophages synthesize interleukin-1 and interleukin-1 mRNA in vitro. J Neurosci Res 21: 391–397

Heumann R, Korshing S, Thoenen H (1987a) Changes of nerve growth factor synthesis in nonneuronal cells in response to sciatic nerve transection. J Cell Biol 104: 1623–1631

Heumann R, Lindholm D, Bandtlow C, Meyer M, Radeke MJ, Misko TP, Shooter E, Thoenen H (1987b) Differential regulation of mRNA encoding nerve growth factor and its receptor in rat sciatic nerve during development, degeneration and regeneration: role of macrophages. Proc Natl Acad Sci USA 84: 8735–8739

Higgins GA, Koh S, Chen KS, Gage FH (1989) Nerve growth factor induces NGF-receptor gene expression in basal forebrain neurons of adult rat. Neurol 3: 241–256

Honegger P, Lenoir D (1982) Nerve growth factor (NGF) stimulation of cholinergic telencephalic neurons in aggregating cell cultures. Dev Brain Res 3: 229–238

Honegger P, DuPasquier P, Tenot M (1986) Cholinergic neurons of fetal rat telencephalon in aggregating cell culture respond to NGF as well as to protein kinase C-activating tumor promoters. Dev Brain Res 29: 217–223

Hosang M, Shooter EM (1985) Molecular characteristics of nerve growth factor receptors on PC12 cells. J Biol Chem 260: 655–662

Johnston MV, McKinney M, Coyle JT (1979) Evidence for a cholinergic projection to neocortex from neurons in basal forebrain. Proc Natl Acad Sci USA 76: 5392–5396

Johnston MV, McKinney M, Coyle JT (1981) Neocortical cholinergic innervation: a description of extrinsic and intrinsic components in the rat. Exp Brain Res 43: 159–172

Johnston MV, Rutkowski JL, Wainer BH, Long JB, Mobley WC (1987) NGF effects on developing forebrain cholinergic neurons are regionally specific. Neurochem Res 12: 985–994

Johnston DG, Silberstein SD, Hanbauer I, Kopin IJ (1972) The role of nerve growth factor in the ramification of sympathetic nerve fibers into the iris in organ culture. J Neurochem 19: 2025–2029

Korsching S, Heumann R, Thoenen H, Hefti F (1986) Cholinergic denervation of the rat hippocampus by fimbrial transection leads to a transient accumulation of nerve growth factor (NGF) without change in mRNA (NGF) content. Neurosci Lett 66: 175–180

Kromer LF (1987) Nerve growth factor treatment after brain injury prevents neuronal death. Science 235: 214–216

Kromer LR, Cornbrooks C (1984) Laminin and a Schwann cell surface antigen present within transplants of cultured CNS cells colocalize with CNS axons regenerating in vivo. Soc Neurosci Abstr 10: 1084

Kromer LF, Bjorklund A, Stenevi U (1981) Regeneration of the septohippocampal pathways in adult rats is promoted by utilizing embryonic hippocampal implants with bridges. Brain Res 210: 173–200

Landmeser L, Pilar G (1978) Interactions between neurons and their targets during in vivo synaptogenesis. Fed Proc 37: 2016–2021

Large TH, Bodary SC, Clegg DO, Weskamp G, Otten U, Reichardt LF (1986) Nerve growth factor gene expression in the developing rat brain. Science 234: 352–355

Larkfors L, Ebendal T, Whittemore SR, Persson H, Hoffer B, Olson L (1987) Decreased level of nerve growth factor (NGF) and its messenger RNA in the aged rat brain. Mol Brain Res 3: 55–60

Letourneau, PC (1978) Chemotaxic response of nerve fiber elongation to nerve growth factor. Dev Biol 66: 183–196

Lewis PR, Shute CCD, Silver A (1967) Confirmation from cholineacetylase of a massive cholinergic innervation to the rat hippocampus. J Physiol (Lond) 191: 215–224

Levi-Montalcini R, Angeletti PU (1963) Essential role of the nerve growth factor on the survival and maintenance of dissociated sensory and sympathetic embronic nerve cells in vitro. Dev Biol 7: 653–659

Levi-Montalcini R, Meyer H, Hamburger V (1954) In vitro experiments on the effects of mouse sarcoma 180 and 37 on the spinal and sympathetic ganglia of the chick embryo. Cancer Res 14: 49–57

Liesi P, Kaakkola S, Dahl D, Vaheri A (1984) Laminin is induced in astrocytes of adult brain injury. EMBO J 683–686

Lindholm D, Heumann R, Meyer M, Thoenen H (1987) Interleukin-1 regulates synthesis of nerve growth factor in non-neuronal cells of rat sciatic nerve. Nature 330: 658–660

Lindsay RM (1979) Adult rat brain astrocytes support survival of both NGF-dependent and NGF-insensitive neurons. Nature 282: 80–82

Loy R, Moore RY (1977) Anomalous innervation of the hippocampal formation by peripheral sympathetic axons following mechanical injury. Exp Neurol 57(2): 645–650

Macdonnell PC, Tolson N, Yu MW, Guroff G (1977) The de novo synthesis of tyrosine hydroxylase in vitro: the effects of nerve growth factor. J Neurochem 28: 843–849

Manthorpe M, Nieto-Sampedro M, Skaper SD, Barbin G, Longo FM, Lewis ER, Cotman CW, Varon S (1983) Neutrotrophic activity in brain wounds of the developing rat. Correlation with implant survival in the wound cavity. Brain Res. 267: 47–56

Martinez HJ, Dreyfus CF, Jonakait GM, Black IB (1985) Nerve growth factor promotes cholinergic development in brain striatal cultures. Proc Natl Acad Sci USA 82: 7777–7781

Martinez HJ, Dreyfus CF, Jonakait GM, Black IB (1987) Nerve growth factor selectively increases cholinergic markers but not neuropeptides in rat basal forebrain in culture. Brain Res 412: 295–301

Max SR, Rohrer H, Otten U, Thoenen H (1978) Nerve growth factor-mediated induction of tyrosine hydroxylase in the rat superior ganglia in vitro. J Biol Chem 253: 8013–8015

Milner TA, Loy R, Amaral DG (1983) An anatomical study of the development of the septohippocampal projection in the rat. Dev Brain Res 8: 343–371

Mobley WC, Rutkowski JL, Tennekoon GI, Buchanan K, Johnston MW (1985) Choline acetyltransferase activity in striatum of neonatal rats increased by nerve growth factor. Science 229: 284–287

Mobley WC, Rutkowski JL, Tennekoon GI, Gemski J, Buchanan K, Johnston MV (1986) Nerve growth factor increases choline acetyltransferase activity in developing basal forebrain neurons. Mol Brain Res 1: 53–62

Montero CN, Hefti F (1988) Rescue of lesioned septal cholinergic neurons by nerve growth factor: specificity and requirement for chronic treatment. J Neurosci 8: 2986–2999

Nieto-Sampedro M, Manthorpe M, Barbin G, Varon S, Cotman CW (1983) Injury-induced neuronotrophic activity in adult rat brain: correlation with survival delayed implants in the wound cavity. J Neurosci 3: 2219–2229

Norrgren G, Ebendal T, Belew M, Jacobsen C-O, Porath J (1980) Release of nerve growth factor by human glial cells in culture. Exp Cell Res 130: 31–39

Olson L, Ebendal T, Seiger A (1979) NGF and anti-NGF: evidence against effects on fiber growth in locus coeruleus from cultures of perinatal CNS tissues. Dev Neurosci 2: 160–176

Olton DS, Walker JA, Gage FH (1978) Hippocampal connections and spatial discrimination. Brain Res 139: 295–308

Otten U, Schwab M, Gagnon C, Thoenen H (1977) Selective induction of tyrosine hydroxylase and dopamine-β-hydroxylase by nerve growth factor: comparison between adrenal medulla and sympathetic ganglia of adult and newborn rats. Brain Res 133: 291–303

Panula P, Revuelta AV, Cheney DL, Wu J-Y, Costa E (1984) An immunohistochemical study on the location of GABAergic neurons in rat septum. J Comp Neurol 222: 69–80

Pearson RCA, Gatter KC, Powell TPS (1983) Retrograde cell degeneration in the basal nucleus of monkey and man. Brain Res 261: 321–326

Peterson GM, Williams LR, Varon S, Gage FH (1987) Loss of GABAergic neurons in the medial septum after fimbria-fornix transection. Neurosci Lett 76: 140–144

Pioro EP, Ribeiro-da-Silva A, Cuello AC (1988) Electron microscopy of nerve growth factor receptor immunoreactivity in cerebellar Purkinje cells of adult rats pretreated with colchicine. Soc Neurosci Abstr 14: 903

Raivich G, Kreutzberg GW (1987) The localisation and distribution of high affinity β-nerve growth factor binding sites in the central nervous system of the adult rat. A light microscopic study using ^{125}I-β nerve growth factor. Neuroscience 20: 23–36

Richardson PM, Verge Isse VMK, Riopelle RJ (1986) Distribution of neuronal receptors for nerve growth factor in the rat. J Neurosci 6: 2312–2321

Riopelle RJ, Verge VMK, Richardson PM (1987) Properties of receptors for nerve growth factor in the mature rat nervous system. Mol Brain Res 3: 45–53

Rohrer H (1985) Nonneuronal cells from chick sympathetic and dorsal root sensory ganglia express catecholamine uptake and receptors for nerve growth factor during development. Dev Biol 111: 95–107

Rohrer H, Barde YA (1982) Presence and disappearance of nerve growth factor receptors on sensory neurons in culture. Dev Biol 89: 309–315

Rohrer H, Sommer I (1983) Simultaneous expression of neuronal and glial properties by chick ciliary ganglion cells during development. J Neurosci 3: 1683–1699

Rudge JS, Manthorpe M, Varon S (1985) The output of neuronotrophic and neurite promoting agents from rat brain astroglial cells: a microculture method for screening potential regulatory molecules. Brain Res 19: 161–172

Schwab ME, Otten U, Agid Y, Thoenen H (1979) Nerve growth factor (NGF) in the rat CNS: absence of specific retrograde axonal transport and tyrosine hydroxylase induction in locus coeruleus an substantia nigra. Brain Res 168: 473–483

Scott SM, Tarris R, Eveleth D, Mansfield H, Weichsel ME, Fisher DA (1984) Bioassay detection of mouse nerve growth factor (mNGF) in the brain of adult mice. J Neurosci Res 6: 653–658

Seiler M, Schwab ME (1984) Specific retrograde transport of nerve growth factor (NGF) from cortex to nucleus basalis in the rat. Brain Res 300: 33–39

Shelton DL, Reichardt LF (1986) Studies on the expression of β-NGF gene in the central nervous system: level and regional distribution of NGF mRNA suggest that NGF functions as a trophic factor for several neuronal populations. Proc Natl Acad Sci USA 83: 2714–2718

Sofroniew MV, Pearson RCA, Eckenstein F, Cuello AC, Powell TPS (1983) Retrograde changes in cholinergic neurons in the basal forebrain of the rat following cortical damage. Brain Res 289: 370–374

Sofroniew MV, Pearson RCA, Powell TPS (1987) The cholinergic nuclei of the basal forebrain of the rat: normal structure, development, and experimentally induced degeneration. Brain Res 411: 310–331

Springer JE, Koh S, Tayrien MW, Loy R (1987) Basal forebrain magnocellular neurons stain for nerve growth factor receptor: correlation with cholinergic cell bodies and effects of axotomy. J Neurosci Res 17: 111–118

Stenevi U, Bjorklund A (1978) Growth of vascular sympathetic axons into the hippocampus after lesions of the septo-hippcampal pathway: a pitfall in brain lesion studies. Neurosci Lett 7: 219–224

Stephens PH, Cuello AC, Sofroniew MV, Pearson RCA, Tagari P (1985) The effect of unilateral decortication upon choline acetyltransferase and glutamate decarboxylase activities in the nucleus basalis and other areas of the rat brain. J Neurochem 45: 1021–1026

Storm-Mathisen J (1974) Choline acetyltransferase and acetylcholinesterase in fascia dentata following lesions of the entorhinal afferent. J Brain Res 80: 119–181

Sutter A, Riopelle RJ, Harris-Warrick RM, Shooter EM (1979) Nerve growth factor receptors. Characterization of two distinct classes of binding sites on chick embryo sensory ganglia cells. J Biol Chem 254: 5972–5982

Taniuchi M, Johnson EM (1985) Characterization of the binding properties and retrograde axonal transport of monoclonal antibody directed against the rat nerve growth factor receptor. J Cell Biol 101: 1100–1106

Taniuchi M, Clark HB, Johnson EM Jr (1986) Induction of nerve growth factor receptor in Schwann Cells after axotomy. Proc Natl Acad Sci USA 83: 4094–4098

Taniuchi M, Clark HB, Schweitzer JB, Johnson EM (1988) Expression of nerve growth factor receptors by Schwann cells of axotomized peripheral nerves: ultrastructural expression by axonal contact and binding proteins. J Neurosci 8: 619–681

Tarris RH, Wieschsel ME Jr, Fisher DA (1986) Synthesis and secretion of a nerve growth-stimulating factor by neonatal mouse astrocyte cells in vitro. Pediatr Res 20: 367–372

Thoenen H, Barde YA (1980) Physiology of nerve growth factor. Physiol Rev 60: 1284–1335

Tuszynski MH, Buzsaki G, Stearns G, Gage FH (1988) Septal cell death following fimbria/fornix transection, and hippocampal cholinergic regeneration following nerve growth factor infusion plus grafting of synthetic and neuronal bridges. Soc Neurosci Abstr 14: 587

Tuszynski MH, Buzsaki G, Gage FH (1990) Nerve growth factor infusions combined with fetal hippocampal grafts enhance reconstruction of the lesioned septohippocampal projection. Neurosci 36: 33–44

Ullrich A, Gray A, Berman C, Dull TJ (1983) Human β-nerve growth factor gene sequence highly homologous to that of mouse. Nature 303: 821–825

Vijayan VK, Cotman (1983) Lysosomal enzyme changes in young and aged control and entorhinal-lesioned rats. Neurobiol Aging 4: 13–23

Wainer BH, Levey AI, Rye DB, Mesulam M, Mufson EJ (1985) Cholinergic and non-cholinergic septohippocampal pathways. Neurosci Lett 54: 45–52

Wainer BH, Hsiang J, Hoffman PC, Heller A, Mobley WC (1986) Nerve growth factor enhances central cholinergic cell survival and fiber proliferation in reaggregate cultures. Neurosci Lett S17

Whitehouse PJ, Price DL, Struble RG, Clark AW, Coyle JT, DeLong MR (1982) Alzheimer's disease and senile dementia: loss of neurons in the basal forebrain. Science 215: 1237–1239

Whittemore SR, Ebendal T, Larkfors L, Olson L, Seiger A, Stromberg I, Persson H (1986) Developmental and regional expression of β-nerve growth factor messenger RNA and protein in the rat central nervous system. Proc Natl Acad Sci USA 83: 817–821

Will B, Hefti F (1985) Behavioral and neurochemical effects of chronic intraventricular injections of nerve growth factor in adult rats with fimbria lesions. Behav Brain Res 17: 17–24

Williams LR, Varon S, Peterson GM, Wictorin K, Fisher W, Bjorklund A, Gage FH (1986) Continuous infusion of nerve growth factor prevents basal forebrain neuronal death after fimbria-fornix transection. Proc Natl Acad Sci USA 83: 9231–9235

Yan Q, Johnson EM (1988) An immunohistochemical study of the nerve growth factor receptor in developing rats. J Neurosci 8: 3481–3498

Yan O, Lampe PE, Clark HB, Johnson EM (1988) Study of NGF receptor in rat posterior pituitary. Soc Neurosci Abstr 14: 902

Zimmerman A, Sutter A (1983) β-nerve growth factor (b-NGF) receptors on glial cells. Cell-cell interactions between neurons and Schwann cells in cultures of chick sensory ganglia. EMBO J 2: 879–885

The Fibroblast Growth Factors: An Emerging Family of Neural Growth Factors

J. A. WAGNER

1 Introduction

The fibroblast growth factors (FGFs) are a family of peptide growth factors that share a number of biochemical and biological properties. The FGFs were originally identified as peptides containing mitogenic activity for fibroblasts and existing at high levels in brain extracts (GOSPODAROWICZ 1974). We now know that the members of the FGF gene family can regulate a number of distinct biological activities. The group of proteins that are known as the FGFs has been

Department of Biological Chemistry and Molecular Pharmacology, Harvard Medical School and Division of Pediatric Oncology, Dana-Farber Cancer Institute, 44 Binney Street, Boston, MA 02115, USA

Current Topics in Microbiology and Immunology, Vol. 165

discovered, rediscovered, and purified under a number of different names reflecting either the tissue of origin or the biological activity that was used to monitor their purification (for a partial list see GOSPODAROWICZ et al. 1987). From the work of a number of laboratories dating from 1983, it as now become apparent that members of the FGF family, in addition to their other biological effects, are potent neural growth factors that have specific effects on the division, differentiation, and survival of specific classes of nerve cells. This review will focus on the effects of the FGFs on nerve cells, but the reader should clearly be aware that these peptides affect a number of other cell types and that a meaningful understanding of their role in development and differentiation can occur only in the context of the constellation of effects under control of these molecules.

2 The FGF Family of Growth Factors

It was originally proposed that FGF was a fragment of myelin basic protein (WESTALL et al. 1978), but this conclusion was shown to be incorrect by THOMAS (1980). Subsequently, the purification of FGF was pursued using a number of different procedures, but the extremely rapid development of this field over the last few years can be traced to the realization that the FGFs have the unusual property of binding strongly to heparin affinity columns (SHING et al. 1984). This discovery soon led to their purification. The novel property of binding to heparin affinity columns at very high salt concentrations also led a number of different groups, working on what were thought to be distinct biological activities, to do experiments demonstrating that they were actually working on the same protein or class of proteins. The names for these proteins included those that reflected their tissue of origin (e.g., retinal derived growth factor) or their biological activities (e.g., astroglial growth factor). In a review of the heparin binding growth factors, which summarizes the evidence that there were only two gene products in the FGF family, LOBB et al. (1986) suggested that the FGFs might more appropriately be called the heparin binding growth factors (HBGFs) both because of the obvious affinity of these growth factors for heparin and because heparin in many cases modulates their biological activity. In this review, we will use the more popular, but more misleading name, FGF.

As suggested by LOBB et al. (1986), the FGFs can be divided into two classes on the basis of their heparin affinities. The class 1 HBGFs were eluted from heparin affinity columns at between 1 and 1.2 *M* salt and had a pI between 5 and 5.8. As a result these growth factors are frequently and most commonly called acidic FGF (aFGF). The class 2 HBGFs also bound heparin affinity columns but were eluted between 1.5 and 2 *M* salt. Because of their basic pI (between 9.2 and 9.8), this class of growth factors is most commonly called the basic FGFs (bFGF). aFGF had been studied under at least nine distinct names including astroglial growth factor type 1, endothelial cell growth factor, retinal derived growth factor, and eye derived growth factor 2. Likewise, bFGF had been studied under at least 23 different names including astroglial growth factor 2, endothelial cell growth factor, hepatoma growth factor, eye derived growth factor 1, pituitary derived growth factor, and chondrocytoma growth factor (for a more complete list and a historical perspective see GOSPODAROWICZ et al. 1987).

On the basis of the sequence information of the peptides and the cross-reactivity between the two cDNAs, a cDNA for both a- and bFGF has been isolated (JAYE et al. 1986, human; ABRAHAM et al. 1986, bovine; SHIMASAKI et al. 1988, rat). This work confirmed that aFGF and bFGF were actually the product of only two distinct genes, one corresponding to bFGF and the other to aFGF. Analysis of these cDNAs suggested that the primary translation product of both a- and bFGF was a 155 amino acid polypeptide containing no signal sequence and that these peptides were about 55% homologous. Furthermore, the genes for a- and bFGF are found on different human chromosomes (chromosome 5 and chromosome 4, respectively, MERGIA et al. 1986). Thus, by 1986, it was clear that many of the endothelial cell growth factors and 3T3 cell growth factors that had been isolated from a number of tissues were products of one of the two genes that encoded aFGF and bFGF. The existence of high molecular weight forms of bFGF (e.g., MOSCATELLI et al. 1987) that contained an N-terminal extension was explained by the demonstration that a mRNA for bFGF could be translated to produce three different species of bFGF (18, 21, and 22.5 kDa), with the larger forms resulting from translational initiation at an in-frame CUG (PRATS et al. 1989). No significant biological differences have yet been reported among the different species of bFGF.

The family of FGFs, however, grew rapidly with the discovery that three distinct oncogene-related proteins are homologous to bFGF: (a) An oncogene has been isolated from Kaposi's sarcoma which encodes a growth factor that is a member of the FGF family (BOVI and BASILICO 1987; BOVI et al. 1987, 1988) and maps to human chromosome 11 (HUEBNER et al. 1988). Unlike a- and bFGF the oncogene isolated from Kaposi's sarcoma (*K-FGF*) expresses a peptide that is cleaved after a signal sequence, glycosylated, and secreted. The media from *K-FGF* transformed cells contained an FGF-like activity, and *K-FGF* transformed 3T3 cells grew in serum free media. The human stomach tumor (*hst*) oncogene (YOSHIDA et al. 1987) is virtually identical to *K-FGF* (BOVI et al. 1988). (2) The *int 2* gene, a site where integration of mouse mammary tumor virus frequently leads to cell transformation and the development of a tumor, encodes a protein that is homologous to the FGFs (SMITH et al. 1988). Normally, the *int 2* gene is thought to be expressed only during early embryonic development, but aberrant expression of this gene in response to activation by mouse mammary tumor virus results in the development of mammary tumors. (3) The *FGF-5* oncogene encodes a protein that is homologous to other members of the FGF gene family and whose expression and secretion can apparently lead to transformation (ZHAN et al. 1988). Thus the family of FGFs includes at least five distinct members, two of which have been isolated as oncogenes and one of which (*int 2*) is clearly related to cell transformation.

3 Regulation of Neural Differentiation and Survival by the FGFs

3.1 PC12 Cells

The first demonstration of an effect of the FGF family on neural differentiation was made by TOGARI who showed that bFGF stimulated neurite outgrowth in PC12 cells (TOGARI et al. 1983, 1985). These papers reported that the initial response of PC12 cells to bFGF was initially comparable to that of NGF; however, in contrast to NGF, the FGF dependent neurites were transient and FGF failed to cause an increase in acetylcholinesterase activity. These initial differences between the activity of bFGF and NGF on PC12 cells have not been substantiated by subsequent work (TOGARI et al. 1985; RYDEL and GREENE 1987; J. WAGNER, unpublished work), and it now appears that NGF and FGF exert essentially identical effects on the differentiation of PC12 cells, including the stimulation of cell-substratum adhesion (SCHUBERT et al. 1987).

The first demonstration that aFGF could also stimulate differentiation of neural cells was also made in the PC12 cell system. The morphological effects of aFGF, purified under the name retina derived growth factor, were indistinguishable at the light microscope level from those of NGF and NGF and aNGF elicited process formation with similar kinetics. Most interestingly, the potency of aFGF was increased 100-fold by the addition of exogenous heparin (WAGNER and D'AMORE 1986). Subsequently, heparin has been shown to strongly potentiate the activity of aFGF in all cases where neural differentiation is effected. Thus, the ability of heparin not only to bind to members of the FGF family, but also to modulate their biological activity is a common theme whose underlying biological meaning must certainly become clearer as the biological role of the FGFs in neural development and differentiation is elucidated.

3.1.1 Heparin Potentiation in PC12 Cells

Presumably the ability of these growth factors to associate with heparin reflects a biologically important association of the growth factor with an endogenously occurring glycosaminoglycan. Modification of FGF action by sulfated glycosaminoglycans is a complex phenomenon, and its physiological meaning and biochemical mechanism will certainly be difficult to unravel. Heparin-like glycosaminoglycans (GAGs) may occur either at the cell surface or as a component of the extracellular matrix. Either of these possibilities would have potentially interesting implications for the control of process outgrowth or nerve regeneration.

Modification of FGF activity is not limited to the effects of heparin on aFGF; other sulfated GAGs including heparan sulfate, dermatan sulfate, and chondrotin sulfate can strongly modify the activity of both aFGF and/or bFGF (DAMON et al. 1988). Since these effects appear to be dependent both on the particular

GAG used and the particular growth factor used, it seems apparent that the effects of these sulfated sugar polymers probably result from an interaction with the growth factor rather than by a nonspecific effect on the cell (DAMON et al. 1988). NEUFELD et al. (1987) confirmed that heparin increased the potency of aFGF but not NGF on PC12 cells but also reported a complex effect of heparin on the biological activity of bFGF. They reported that heparin stimulated the biological activity of low concentrations of bFGF while it inhibited the biological activity of high concentrations of bFGF. How those experiments can be reconciled with the observations of DAMON et al. (1988) is not readily apparent but may be due to methodological differences. SCHUBERT et al. (1987) made the intriguing observation that bFGF, when absorbed to an extracellular matrix which contained GAGs, caused the formation of aggregates or circles of PC12 cells.

The ability of heparin to potentiate the action of aFGF on PC12 cells in culture may, to a large extent, be explained by the ability of heparin to extend the biological half-life of this growth factor (DAMON et al. 1989) and probably reflects the ability of heparin to stabilize a- and bFGF in vitro both to heat inactivation and protease attack (GOSPODAROWICZ and CHENG 1986). The varying potencies of aFGF and bFGF may be at least partially due to a differential sensitivity to proteases (LOBB 1988; DAMON et al. 1989). The 100-fold inhibition of bFGF activity by heparan sulfate (DAMON et al. 1989) must have a different explanation, possibly as simple as a masking of the biological activity of FGF by binding to the growth factor. Since the biological potency of the FGFs is strongly modified by the presence of GAGs and proteases, a full understanding of the role of the FGFs will undoubtedly require an understanding not only of the production of these factors, but also an understanding of the ways proteases and GAGs can modify expression of their biological activity.

3.2 Chromaffin Cells

Chromaffin cells, which share many properties with PC12 cells including the ability to respond to NGF both by dividing and by differentiating into a sympathetic nerve-like cells, also respond to bFGF (STEMPLE et al. 1988) and aFGF (CLAUDE et al. 1988). The study of the FGFs in chromaffin cells, which, in contrast to PC12 cells, are not an established cell line, has revealed an extremely interesting difference between the biological activity of NGF and the FGFs. While NGF is both a differentiation factor and a long-term survival factor for the sympathetic nerve-like cells that are derived from chromaffin cells, both aFGF and bFGF appear to be able to initiate their differentiation and stimulate their proliferation without acting as long-term survival factors (STEMPLE et al. 1988; CLAUDE et al. 1988). In fact, the differentiated cells that result from the action of bFGF apparently become dependent on NGF for continued survival (STEMPLE et al. 1988). Dexamethasone antagonizes the neurite outgrowth elicited by NGF, aFGF, and bFGF; while dexamethasone inhibits the NGF-induced, but not the

FGF-induced, proliferation of these cells (Stemple et al. 1988; Claude et al. 1988). This points to another important difference between NGF and FGF dependent pathways and leads to the suggestion that NGF and the FGFs may serve different functions during development. For example, in the adrenal gland, bFGF or aFGF may be the stimulus that allows amplification of the chromaffin cell population, while the presence of glucocorticoids prevents neural differentiation. Likewise, the FGFs could conceivably be involved in the initial differentiation and neurite formation of sympathoblasts, while NGF may play the role of a maturation factor or a survival factor.

3.3 Peripheral Nerves

Unlike NGF, the FGFs do not appear to have the ability to act as survival factors for sympathetic nerve cells (Unsicker et al. 1987). Both aFGF and bFGF, however, stimulate survival and neurite outgrowth from chick ciliary ganglion cells, which are parasympathetic, and the effects of aFGF are strongly potentiated by heparin (Unsicker et al. 1987; Schubert et al. 1987). Thus, in vitro, aFGF and bFGF have properties that strongly suggest they are neurotrophic factors and neurite extension factors for cells from the peripheral nervous systems but in vivo evidence for this assertion is still required. The FGFs and NGF appear to influence an overlapping but distinct population of cells, and they appear to act by biochemical mechanisms that are at least partially distinct.

3.4 Nerves from the Central Nervous System

The FGFs, which are present in high concentrations in the central nervous system, also stimulate the survival and/or differentiation of a number of mixed populations of nerves from different regions of the CNS. bFGF is both a survival factor and a neurite extension factor for cortical neurons and purified hippocampal neurons, but not cerebellar neurons (Morrison et al. 1986; Walicke et al. 1986; Morrison et al. 1988); however, in the case of hippocampal cells, it is clear that factors other than FGF must also be required for long-term survival. Although bFGF obviously has effects on non-neural cells that are present in these cultures, it is most likely that the effects of FGF are directly on the population of neural cells (Walicke and Baird 1988). A survey of the effects of aFGF and bFGF on several areas of the brain by Walicke (1988) has shown that both aFGF and bFGF are potent neurotrophic factors for a variety of populations of cells from many different regions of the brain. In particular, bFGF increased survival of cells from the hippocampus, entorhinal cortex, frontal cortex, parietal cortex, occipital cortex, striatum, septum, and thalamus, but not from the subiculum. Interestingly, aFGF supported a smaller number of neurons from each area where bFGF had an effect, and, in contrast to bFGF, aFGF increased survival of a population of nerves from the subiculum. This data suggests that aFGF and bFGF influence nonidentical populations of nerve cells and may act through different receptors (WALICKE 1988a).

bFGF also enhanced the survival of a population of spinal cord neurons including some motor neurons (UNSICKER et al. 1987). bFGF also appears to modestly stimulate the proliferation of neuronal precursor cells (GENSBURGER et al. 1987) and enhances both the survival and morphological differentiation of a population of fetal rat brain cells including oligodendrocytes, astrocytes, and some neurons (ECCLESTON et al. 1985).

FGFs also have striking effects on several definitively identified neural cell types. bFGF maintains the viability and promotes the differentiation of a purified population of granule cells (HATTEN et al. 1988). Astroglial cells can maintain granule neurons in culture in the absence of exogenous bFGF, and this effect appears to be mediated, at least in part, by astroglial derived bFGF (HATTEN et al. 1988). aFGF also enhanced the regeneration of retinal ganglion cells (LIPTON et al. 1988) at a time when there was no obvious effect on survival; however, future experiments will be required to assess the role of aFGF in long-term survival of these cells. The biological potency of aFGF on retinal ganglion cells was markedly potentiated by heparin and, interestingly, there was no apparent effect of bFGF on these cells. Retinal ganglion cells stand as one of two examples of a neural cell type that responds to aFGF but not bFGF (with the aFGF responsive population in the subiculum being the other [WALICKE 1988a, b]). Statistical analysis demonstrates that aFGF can contribute to both neurite elongation and initiation by retinal ganglion cells. Finally, aFGF enhances both the survival and the expression of opsin by retinal photoreceptor cells (HICKS and COURTOIS 1988), and FGF increase the synthesis of prolactin by GH3 cells, a pituitary derived tumor line (MORMEDE and BAIRD 1988).

Thus, in a rapidly increasing number of cases, either defined cell types or populations of cells that include multiple types of neurons from either the central or peripheral nervous system have been shown to respond to aFGF and/or bFGF with enhanced survival and/or neurite formation. Since the FGFs can clearly act as survival factors, it remains possible that their effects on neurite formation by these cell types may sometimes be a secondary effect of their enhancement of survival. Thus, it will be prudent to be aware of these possible complications in designing future experiments. Nevertheless, the fact that both aFGF and bFGF have been shown to stimulate the differentiation of a transformed cell line that does not depend on FGF for survival (PC12 cells) certainly supports the idea that these growth factors may have effects on the expression of differentiated characteristics beyond their effects on cell survival.

3.5 Astrocytes and Oligodendrocytes

Basic FGF has long been known to be an essential component of a chemically defined media for astrocytes (MORRISON and DE VELLIS 1981), and bFGF can stimulate quiescent astrocytes to re-enter the cell cycle (KNISS and BURRY 1988).

PETTMANN et al. (1985) purified two astrocyte mitogens from brain (astroglial growth factor 1 and astroglial growth factor 2) which correspond to aFGF and bFGF, respectively. In addition to stimulating the division of astrocytes, a- and bFGF stimulate the acquisition of a star-like morphology that is characteristic of mature astrocytes and increase the expression of glial fibrillary acidic protein, a marker of mature astrocytes (PERRAUD et al. 1988; MORRISON et al. 1985). bFGF stimulates the synthesis of plasminogen activator (ROGISTER et al. 1988). The responses of astrocytes to aFGF were somewhat different from the responses to bFGF. At the ultrastructural level, aFGF was more effective at increasing the expression of neurofilaments while bFGF had a greater effect on the abundance of microtubules, leading to the suggestion that aFGF and bFGF may act through distinct signal transduction systems in astrocytes (PERRAUD et al. 1988). It is unclear if the effects of the FGFs on the expression of the differential phenotype by astrocytes is due to an FGF requirement for their survival in culture or if either of the FGFs have a direct effect on the differentiation of these cells; thus, experiments that distinguish the effects of aFGF and bFGF on survival, division, and differentiation will be of great interest. The astrocytoma line, U87-MG, expresses bFGF and the expression of the gene is regulated by serum, by the activity of protein kinase C and, apparently, by cell-cell contact, suggesting an autocrine role for the growth factor (MURPHY et al. 1988a, b). Such a regulatory pattern, if found in normal astrocytes, would explain the relatively low level of FGF expression normally observed, and it would be consistent with the suggestion that bFGF is expressed in response to tissue damage and that its expression is involved in coordinating tissue repair or remodeling after injury. FGF also stimulates the primitive neural mouse cell line F7 to differentiate into an oligodendrocyte-like cell type, as indicated by changes in cell morphology and the expression of carbonic anhydrase and myelin basic protein, supporting the idea that the FGFs may play a role in producing the oligodendrocyte cell lineage (DE VITRY et al. 1983).

4 Effects of the FGFs on Nerve Regeneration and Cell Survival In Vivo

Acidic FGF stimulated an increase in the number of myelinated axons that regenerated across a transected sciatic nerve. This increase was due largely to an increase in the number of sensory, and to some extent, motor neurons that extended axons into the distal stump (CORDEIRO et al. 1989). This increase in nerve regeneration may be due, at least in part, to the ability of FGF to protect adult sensory nerves from lesion-induced cell death (OTTO et al. 1987). Likewise, administration of bFGF has been shown to enhance the survival of cholinergic neurons in the medial septum and diagonal band of Broca after transection of their axons (ANDERSON et al. 1988). Thus, the FGFs are clearly capable of acting as survival factors in vivo substantiating the observations that have been made in culture and strengthening the hypothesis that they are important neurotrophic factors that act to maintain an appropriate number of mature neurons in vivo.

5 Expression of aFGF and bFGF in Neural Tissue

Basic FGF has been isolated from a broad range of cell types and tissues while aFGF has been isolated only in nerve and a few other tissue types (see GOSPODAROWICZ et al. 1987 for a review). On the basis of immunochemical localizations bFGF is thought to be found (although it is not necessarily synthesized) in neurons (FINKLESTEIN et al. 1988; PETTMANN et al. 1986; JANET et al. 1987). The expression of bFGF and aFGF can be dramatically increased after focal brain injury (FINKLESTEIN et al. 1988; NIETO-SAMPEDRO et al. 1988), and bFGF appears to be localized at the border of the lesion in a cell that resembles a reactive astrocyte (FINKLESTEIN et al. 1988). SHIMASAKI et al. (1988) have shown that bFGF mRNA is expressed in cortex and hypothalamus, but is below detectable levels in a number of non-neural tissues from which bFGF can be purified. This suggests either that the high level of bFGF found in these non-neural tissues is due to storage of the mitogen or that bFGF is synthesized only under specific circumstances by these tissues. The expression of bFGF in brain, where cell division and neovascularization are rare, is also of interest because it highlights the idea that one of bFGF's major roles may be to regulate neural function. bFGF is also produced in several classes of cells in culture including retinal pigment epithelia (SCHWEIGERER et al. 1987), pituitary cells (FERRARA et al. 1987), epithelial cells (MOSCATELLI et al. 1986), vascular smooth muscle cells (WINKLES et al. 1987), and astrocytes (discussed above); however, the observation that bFGF production is regulated in astrocytes by their position in the cell cycle emphasizes that observations made in cell culture should not be carelessly extrapolated to in vivo situations. aFGF is expressed in brain and retina (ALTERIO et al. 1988), but the apparent need to purify poly A^+ RNA to identify the transcripts testifies to its lack of abundance. It is clearly of importance to determine which cells in the nervous system express the FGFs, how that expression changes with development, and how environmental factors (such as injury, change of synaptic input, anoxia, or cell death) regulate its expression.

6 Effects of the FGFs on Non-Neural Cells

In contrast to NGF, which was originally described as a neurotrophic factor and only later came to be recgonized as a molecule that could influence the behavior of non-neuronal cells, the study of FGFs' effects on nerves and other cells in the nervous system takes place in the context of the realization that FGFs have a number of significant effects on other cell types. Although it is impossible for this review to describe these biological effects in any detail, a brief list of the most significant observations seems essential.

6.1 The FGFs as Oncogenes

Given that the original assays that led to the purification and characterization of the FGFs were based on the ability of these molecules to act as mitogens for a number of cell types, most prominently fibroblasts and endothelial cells, it was initially suprising that FGFs or FGF-like gene products were never isolated as oncogenes. This apparent paradox has been resolved by the isolation of at least three members of the FGF family (*K-FGF/hst, int2,* and *FGF-5*) as oncogenes (Bovi et al. 1987; Smith et al. 1988; Yoshida et al. 1987; Zhan et al. 1988). The failure of the focus-forming assays to isolate genes corresponding to aFGF and bFGF may be partially explained by the fact that the expression or overexpression of these genes can confer a partially transformed, but not a fully transformed, phenotype (Jaye et al. 1988; Neufeld et al. 1988; Sasada et al. 1988). bFGF itself may be unable to contribute strongly to tumor formation because it is not released efficiently into the medium in its native state; however, modification of the gene by the addition of a signal sequence leads to a form of bFGF that can induce a transformed phenotype, including the formation of tumors (Rogelj et al. 1988). It is of some interest that antibodies of bFGF can actually inhibit the growth of pulmonary endothelial cells, suggesting an autocrine role for bFGF (Sakaguchi et al. 1988). Thus, as would be expected from their biological properties, the expression of altered FGFs or the inappropriate expression of the FGFs can result in tumor formation. Likewise, the appreciation of the potential of the FGFs to contribute to the transformed phenotype will undoubtedly lead to a more detailed definition of their role in tumor progression in the near future (for a recent review see Thomas 1988). A role for the FGFs in the development of neural tumors has not yet been established.

It should not be forgotten that the FGFs may also contribute to tumor progression by virtue of their ability to act as endothelial cell mitogens and angiogenic factors leading to increased vascularization, an essential step in the development of a clinically significant tumor (for a review see Folkman and Klagsbrun 1987).

6.2 The FGFs as Inducing Factors During Embryogenesis

The FGFs have also been implicated in the regulation of early embryonic development in *Xenopus*. Slack et al. (1987) reported that bFGF can mimic the ventrovegetal signal that is essential for induction of mesodermal tissue. Furthermore, the natural ventrovegetal signal can be blocked by heparin, which could bind secreted FGF and reduce its concentration, further suggesting that a HBGF such as bFGF may be the inducer. Of course, the existence of a family of FGF-like molecules makes any such assignment premature. Nevertheless, the data is extremely suggestive and exciting and will certainly help to more clearly identify the essential endogenous signals and the sights of their synthesis. A mRNA encoding a protein with very strong homology to bFGF has also been isolated from *Xenopus*, and the FGFs, in conjunction with other peptide growth factors, have been shown to strongly stimulate the expression of mesodermal markers (KIMELMAN and KIRSCHNER 1987; KIMELMAN et al. 1988). *Int-2* is expressed in a number of tissues during embryonic development including the neuroepithelium, suggesting that this member of the FGF family may play a role in the early development of the nervous system (WILKINSON et al. 1988; JAKOBOVITS et al. 1986).

6.3 The FGFs as Mitogens

The FGFs were originally purified and identified by their ability to act as mitogens on a number of cell types, most prominently fibroblasts, endothelial cells, astrocytes, and oligodendrocytes; more recently, they have been shown to be mitogens for neuroblasts, keratinocytes, bovine epithelial lens cells, osteoblasts, and melanocytes (e.g., LOBB et al. 1986; GOSPODAROWICZ et al. 1986; PETTMANN et al. 1985; O'KEEFE et al. 1988; RISTOW and MESSMER 1988; RODAN et al. 1987; GLOBUS et al. 1988; CANALIS et al. 1988; MOENNER et al. 1986; HALABAN et al. 1988a, b; WU et al. 1988).

6.4 The FGFs as Differentiation Factors in Non-Neural Cells

The FGFs regulate the differentiation or expression of the differentiated phenotype of a number of non-neural cells. For example: (1) The FGFs are implicated at several steps in muscle differentiation. FGF deprivation induces the formation of post-mitotic cells from proliferating myoblasts and FGF addition both delays the onset of myotube formation and can actually reverse the expression of muscle specific RNAs by a mechanism that is distinguishable from the effect of FGF on myoblast proliferation. The muscle colony forming cells (chick stage 23 to 27) require FGF to differentiate, while loss of the FGF receptor is correlated with the generation of a post-mitotic state (SEED and HAUSCHKA 1988; SPITZ et al. 1986; HAUSCHKA et al. 1987; KARDAMI et al. 1988). (2) FGF plays an important role in the maturation of ovarian granulosa cells by blocking follicle stimulating hormone dependent increases in aromatase activity (BAIRD and HSUEH 1986; BISWAS et al. 1988; ADASHI et al. 1988). (3) FGF has also been reported to stimulate estrogen production by Leydig cells (RAESIDE et al. 1988). (4) Basic FGF can inhibit the differentiation of adipoblasts to adipocytes (NAVRE and RINGOLD 1988).

7 Regulation of aFGF and bFGF Activity by Heparin-Like Molecules and Proteases

One of the distinguishing characteristics of the FGF family is that they all appear to retain the ability to bind strongly to heparin in such way that the binding can be disrupted only at extremely high salt concentrations. This strong physical interaction of the FGFs with heparin is reflected by the ability of heparin and heparin-like molecules to strongly modify the biological activity of these growth factors. I know of no other case in which a well-defined growth factor has been demonstrated to associate so strongly with a known component of the extracellular matrix; and there is a clear interaction between cells and components of the extracellular matrix during neurite formation, (e.g., TOMASELLI et al. 1987; EDGAR et al. 1984; LANDER et al. 1982). Thus, it is worth considering both the effects of heparin on the biological activities of these molecules and the possible biological meaning of the retention of this strong heparin avidity in a family of growth factors. Certainly it is unlikely that the strong association of the FGFs with heparin and/or heparin-like molecules is purely an accident; more likely, the ability to bind to sulfated GAGs is a product of natural selection and this binding plays an important role in their'normal biological activity. The nature of this role is still far from clear; however, a number of possibilities deserve further analysis and experimental attention.

Heparin potentiates the biological activity of aFGF in PC12 cells, adrenal chromaffin cells, retinal ganglion cells, and ciliary ganglion (WAGNER and D'AMORE 1986; DAMON et al. 1988; CLAUDE et al. 1988; LIPTON et al. 1988; UNSICKER et al. 1987; NEUFELD et al. 1987). This potentiation is so striking (100-fold) that in the absence of exogenous heparin only unphysiologically large concentrations of aFGF are effective in stimulating neural differentiation and/or survival. Heparin is not the only sulfated GAG that can substantively potentiate aFGF; for example, dermatan sulfate is nearly as effective as heparin in potentiating neurite extension by PC12 cells (DAMON et al. 1988). Furthermore, the effect of heparin and heparin-like molecules appears to be specific for aFGFs, and a similar pattern of potentiation is not observed with other peptide growth factors (NGF and/or bFGF), suggesting that a specific interaction between the growth factor and the GAG or the growth factor, the GAG, and the cell is important for potentiation (DAMON et al. 1988).

The effect of heparin-like molecules on the biological activity of aFGF is not unique to nerve cells, but appears to be a common feature of this growth factor's effect on many cell types. Heparin strongly potentiates the activity of aFGF on endothelial cells and has been an essential component of effort to use aFGF to allow their long-term cultivation (THORNTON et al. 1983; GIMENEZ-GALLEGO et al. 1986). Heparin also potentiates the mitogenic effect of aFGF on several types of fibroblasts and bovine epithelial lens cells (RYBAK et al. 1988; LOBB et al. 1986; UHLRICH et al. 1986).

Heparin and related sulfated GAGs can modestly (5-fold) potentiate the biological activity of bFGF on PC12 cells (DAMON et al. 1988). The ability of various sulfated GAGs to potentiate bFGF is not correlated with the potency of the sulfated GAGs to potentiate aFGF, again suggesting that there are specific interactions between the GAG and the growth factor that result in the modification of biological activity. Heparan sulfate which is closely related structurally to heparin is capable of strong inhibiting (100-fold) the biological activity bFGF (DAMON et al. 1988). Thus the interaction between sulfated GAGs and the FGFs have the capacity either to inhibit or to potentiate the biological activity of the growth factors. Likewise, heparin has been used to inhibit the natural induction of mesoderm in tissues derived from *Xenopus* embryos (SLACK et al. 1987), an observation that has been interpreted to suggest that heparin or heparin-like molecules may be able to modify the biological potency of FGFs, possibly by sequesting the FGF into a bound and inactive form. Pretreatment of substrates with heparin appears to potentiate the ability of bFGF to stimulate neurite outgrowth from hippocampal neurons (WALICKE et al. 1986), and the ability of various GAGs to potentiate FGF activity is directly correlated with their ability to bind bFGF to GAG-coated plates (WALICKE 1988b).

As sulfated GAGs seem to have the ability both to potentiate and to inhibit the biological activities of the FGFs, the explanation for these important phenomena will undoubtedly be complex. Certainly sequestration of FGFs into an inactive or unavailable form by binding to GAGs is an important part of any proposed mechanisms, but it is difficulte to see how this type of mechanism might explain potentiation. Potentiation is more likely to be explained by the ability of GAGs to bind to the FGF and either protect the FGF from inactivation (e.g., proteolytic degradation) or stabilize it in a more active conformation. It has been known for some time that heparin can protect bFGF and aFGF from proteolytic inactivation (GOSPODAROWICZ and CHENG 1986). This work suggests that the binding of GAGs to FGFs modifies their tertiary structure into one that is less accessible to proteases; however, such data cannot be extrapolated to explain the biological effects of heparin on growth factor activity without caution. Recent work (DAMON et al. 1989) has clearly demonstrated that one of the key ways that heparin potentiates the biological activity of aFGF is by extending its biological half-life. This hypothesis was supported by the observation that a specific protease inhibitor (aprotinin) could potentiate the biological activity of aFGF and that heparin potentiation could not be observed under conditions where aFGF was re-added frequently to the media (DAMON et al. 1989). It is interesting to speculate that the existence of several members of the FGF family may be partially due to the need to have specific mechanisms for inactivating these growth factors under different circumstances. For example, thrombin can specifically inactivate aFGF but not bFGF (LOBB 1988). It has also been suggested that protection from proteolytic inactivation may explain heparin potentiation of aFGF dependent mitogenicity on endothelial cells (ROSENGART et al. 1988).

Thus, heparin and related GAGs may modulate the biological activity of the FGF family by a number of distinct mechanisms. The simplest of these mechanisms would include the ability of FGF to bind to these molecules and be sequestered or protected from proteolytic inactivation; however, more complex mechanisms cannot be ruled out at present. For example, a dual recognition of FGF and associated heparin by a complex cell surface receptor might result in a stronger activation of cell's surface receptors, or the change in the structure of the FGF peptide might modify the type of cell surface receptor that is recognized. The most intriguing question that needs to be answered is the biological role served by the association of the FGFs with heparin. Is this purely a means of extending their biological activity or is it potentially a method of restricting the availability of these biologically potent trophic molecules to a local area? bFGF has been shown to bind to the heparan sulfate that is present in the basement membrane (VLODAVSKY et al. 1987; BAIRD and LING 1987; JEANNY et al. 1987; FOLKMAN et al. 1988; VIGINY et al. 1988), and endothelial cells have been shown to produce a heparan sulfate proteoglycan that could bind to bFGF and protect it from proteases (SAKSELA et al. 1988). BAIRD et al. (1988) have used synthetic peptides to explore the regions of the bFGF peptide that are required for interaction with heparin and binding to the biological receptor. It is reasonable to speculate that the FGFs may be stored in the basement membrane where they could be protected from extracellular proteases or presented to cells in the context of the basement membrane. Might the FGFs be available to cells only in certain circumstances (e.g., during degradation of the extracellular matrix) and thus be present in the body but unavailable to cells because of their unique location? Could bound FGF be serving as a specific marker to stimulate the extension of neuronal processes or migration of cells along specific pathways? These are questions that are truly unique to the FGFs and their answers will almost certainly provide a profound insight into the mechanisms used by organisms to control morphogenesis, differentiation, and repair of nerves as well as other cell types.

8 Mechanism of Action of the FGFs

The study of the way FGF interacts with its receptor and how occupation of this receptor may regulate nerve survival, neurite extension, or the myriad of other responses induced by FGF is, as might be expected, in its infancy, but the work that has been done has begun to give us an insight into FGF-regulated cell behavior. Before reviewing the literature in this field, consideration of some of the problems that will confront us in understanding the FGF receptor and the biochemical mechanism of action of FGF will be beneficial.

First, as we have already noted, the FGFs are a gene family and there may be a heterogeneity within the products of a single gene because of modifications in proteolytic processing (KLAGSBRUN et al. 1987) and/or mRNA splicing. This immediately suggests the possibility that multiple receptors may exist and that these receptors may recognize different members of the family with different affinities. A precedent for this type of complexity clearly exists in other peptide hormones, such as insulin and the insulin-like growth factors or PDGF (HELDIN et al. 1988). Thus, the assignment of a particular receptor to a particular member of the FGF family should be done with caution. This assignment of a particular FGF to any particular physiological process must require evidence from a number of distinct areas, including not only the ability of a particular member of the family to regulate the response being studied, but also evidence that the family member is present in the appropriate tissue and secreted or released at the appropriate time for the cell to respond. Thus, particular attention to dose-response curves, binding curves, in situ localization studies, and, eventually, the analysis of mutants may prove to be important in this field.

Second, it is not at all clear that different classes of differentiated cells will share the same receptors or that the same receptor will be expressed on any cell type at all times in development. Third, the fact that FGFs all bind tightly to GAGs introduces a problem that is unique to the study of FGFs. It is important to guard against potential artifacts resulting from the binding of the FGFs to endogenous GAGs. On the other hand, it is completely possible that associaton of the FGFs with endogenous GAGs may be important, or even essential, for receptor function. A dual recognition of the FGF molecule (by both a classical receptor and a cell surface sugar) may have a number of important effects including modulating the effect of the FGF on its receptor, prolonging the half-life of the growth factor-receptor complex, modulating the efficacy of the FGF, etc. Finally, it is important to ascertain whether binding of the FGFs to GAGs that are present in the extracellular matrix have a biologically important function. Such binding may: (1) stabilize the FGF to proteases or other enzymes, (2) present the growth factor to appropriate cells while precluding its recognition by inappropriate cells, or (3) result in the exposure of active FGF during physiological processes of matrix degradation. The number of models in which various GAGs might be involved in the action of FGF is really limited only by the imagination of the individual investigator, as the amount of data that really bears on this issue is still quite small; however, the retention of specific binding to heparin and other sulfated sugars through evolution clearly points to an important role for this interaction and an explanation certainly must be obtained before we can claim to understand how the FGFs regulate cell division, differentiation, and determination.

The FGF receptor has been investigated in a number of cell types using equilibrium binding and cross-linking studies, and there appears to be both a high affinity $K_d = 10^{-11}\,M$) and a low affinity ($K_d = 10^{-9}\,M$) receptor. Cross-linking studies have identified a receptor with a molecular mass in the range of 125–170 kDa depending on the particular cell type that was studied. FGF receptors have been identified in endothelial cells, muscle cells, fibroblasts, PC12 cells, hepatoma lines, epithelial cell lines, baby hamster kidney (BHK) cells, and primary cultures from several regions of the brain that are enriched in neurons (NEUFELD and GOSPODAROWICZ 1988; MOSCATELLI 1987, 1988; FARIEFEL 1986, endothelial cells; OLWIN and HAUSCHKA 1986, muscle; COURTY et al. 1988, brain; NEUFELD et al. 1987, PC12 cells; MOENNER et al. 1986, bovine lens epithelial cells;

KAN et al. 1988, hepatoma; NEUFELD and GOSPODAROWICZ 1985, BHK cells; andWALICKE et al. 1989, brain). In addition to the high molecular weight FGF receptor, neuron enriched cultures from the brain contained an 85 kDa binding protein. The apparent affinity of the FGF receptor in these cultures is significantly lower than that found in other cell types (WALICKE et al. 1989), supporting the possibility that the form of the FGF receptor found in the brain may be distinct from the form found in mesenchymal cells. The demonstration that several classes of neural cells and membranes that are prepared from the brain have high affinity receptors for FGF supports the assertion that these peptides are indeed neural growth factors.

MOSCATELLI (1987) demonstrated that the high affinity binding site for bFGF was acid sensitive and seemed to be associated with the biological receptor whose occupation led to cell division and plasminogen activator production in endothelial cells. The low affinity site was salt sensitive and appeared to be an association of the growth factor with heparin-like molecules on the cell. BHK cells are a useful experimental system for the study of FGF because they have an unusually high number of receptors (1.2×10^5) per cell and a strikingly low concentration of low affinity binding sites that appear to be saturable (NEUFELD and GOSPODAROWICZ 1985). The effect of heparin on binding to the high affinity receptor is complex. MOSCATELLI (1988) has reported that heparin competes strongly for binding of bFGF to low affinity sites but does not compete strongly for binding to the high affinity sites. NEUFELD and GOSPODAROWICZ (1986), however, have reported that heparin strongly inhibits cross-linking of FGF to the high affinity receptor in BHK and PC12 cells, while heparin did not inhibit the cross-linking of aFGF to its receptor. Nevertheless, aFGF and bFGF were reported to compete for the same receptor. SCHREIBER et al. (1985) have suggested that heparin actually increases the affinity of aFGF for its receptor in endothelial cells. There is no obvious explanation for these apparently conflicting sets of data, but as a more detailed biochemistry of the receptor(s) becomes available these observations should be clarified. After exposure to aFGF or bFGF, the FGF receptor is internalized resulting in a population of occupied receptors in which FGF is no longer available at the cell surface. In some cases the FGF that is internalized appears to be degraded (FRIESEL and MACIAG 1988; MOSCATELLI 1988).

The FGF receptor has been characterized in a number of other biochemical ways. In BHK cells, it is a glycoprotein with a core protein of between 100 and 125 kDa that is glycosylated to a high mannose form of 115–140 kDa which is processed to a mature species of 130–150 kDa. It binds strongly to wheat germ agglutinin, which inhibits the binding of FGF to the receptor. Furthermore, the core glycoprotein is unable to recognize and bind the FGFs suggesting that the *N*-acylglucosamine residues are functionally significant components of the receptor (FEIGE and BAIRD 1988). aFGF also stimulates the phosphorylation of a 135 kDa protein on tyrosine residues, and this protein can be cross-linked to the growth factor suggesting either that the FGF receptor is itself, or is tightly associated with, a protein tyrosine kinase (HUANG and HUANG 1986; COUGHLIN et al. 1988). The FGF receptor has been purified using affinity chromatography on both wheat germ agglutinin and bFGF affinity columns. The purified receptor consisted of two species, one with a molecular mass of 110 kDa and a second with a molecular mass around 140 kDa although the latter appeared to be by far the predominant species eluted at higher concentrations of heparin (IMAMURA et al. 1988). Both aFGF and bFGF have been shown to interact with two FGF receptors, but on the basis of competition studies it appears that aFGF binds preferentially to the 125 kDa species, while bFGF binds preferentially to the 145 kDa species (NEUFELD and GOSPODAROWICZ 1986).

The second messenger systems that are regulated by the occupation of the FGF receptor(s) have not yet been completely defined, but accumulating evidence suggests the possible involvement of several of the classical second messenger pathways. As noted above, the FGF receptor is associated with a tyrosine kinase (HUANG and HUANG 1986; COUGHLIN et al. 1988). Occupation of the FGF receptor leads to phosphoinositide turnover and activation of protein kinase C (TSUDA et al. 1985; KAIBUCHI et al. 1986; MOSCATELLI 1988), although there is also evidence that this second messenger system is not essential for FGF action in some cases (MAGNALDO et al. 1986). aFGF has also been reported to activate adenylate cyclase in smooth muscle cells (MIOH and CHEN 1987), lead to activation of the S6 kinase (PELECH et al. 1986), and regulate ion flux (HALPERIN and LOBB 1987). BOUCHE et al. (1987) reported that FGF is translocated to the nucleolus and that exogenous FGF added to isolated nucleoli can stimulate rRNA transcription, although it is unclear if they were studying a specific effect of the growth factor. In PC12 cells and in fibroblasts, FGF rapidly regulates gene expression (GREENBERG et al. 1985; CHO et al. 1989; LEONARD et al. 1987; KAIBUCHI et al. 1986), but the pathways responsible for this effect have not yet been established. In PC12 cells, the protein kinase inhibitor k-252a can block the biological responses to NGF but not FGF. Thus, K-252a could be a useful pharmacological agent in mechanistic and physiological studies of these two growth factors (KOIZUMI et al. 1988; CHO et al. 1989; HASHIMOTO 1988). Given the pleiotropic response induced by the FGFs, the signaling pathway used by the FGFs would be expected to be complex. Moreover, because of the diversity of responsive cells, there may be variations in the second messenger systems used by different classes of differentiated cell types. It is also reasonable to expect that there may be an interaction among the signal transduction systems regulated by the FGFs and other neurotrophic factors, like NGF. A physiological understanding of the role of the FGFs will require an understanding of this relationship.

9 The Growing Puzzle of the FGFs: More Questions than Answers

In contrast to NGF which has been studied for 50 years, the effects of the FGFs on neural cells have become apparent only in the last few years; yet, because of modern methods and the current appreciation of the ways one might expect a neural growth factor to work, the rate of progress has been remarkable. The work reviewed above has provided the fundamental information that is needed to phrase the really significant questions. The study of the role of the FGF family in neural development promises to be one of great importance and great complexity. There is persuasive evidence that members of this family can stimulate neurite formation, promote cell survival, and regulate differentiation in specific classes of nerve cells; however, there are at least five major members of the FGF family, and members of this family affect a diverse set of neural and non-neural cells. Does this reflect some underlying coordination of developmental processes, or is it the use of the same signalling molecule in different circumstances for different purposes, much as the same set of neurotransmitters are used for many different functions? For example, could the FGFs be involved in coordinating vascularization and innervation during development or wound healing? Are there a family of FGF receptors that preferentially recognize specific members of the FGF family, or are there specific FGF receptors that are linked to specific second messenger systems? Why has the ability to bind to GAGs been conserved in this family, and which of the GAGs found in the organism are important for the regulation of FGF activity? What cells produce the FGFs? How are the FGFs stored? How is the release of aFGF and bFGF which lack the classic leader sequence regulated? Is release of aFGF and bFGF after cell lysis a biologically important pathway or is there another way these molecules are released? What cells respond to the FGFs during development? How are the expression of the FGFs regulated? What are the roles of the FGFs that are preferentially expressed during embryogenesis? What classes of nerve cells respond to the FGFs? Do the FGFs regulate cell survival or stimulate neurite outgrowth in vivo? Are some cells transiently responsive to particular FGFs? Would the failure of the FGF dependent pathways lead to degenerative diseases of the nervous system or other pathological conditions? Does the presence of heparan sulfate proteoglycans in the amyloid structures that develop in Alzheimer's disease implicate the FGFs in the development of this disease? Could therapies using the FGFs or pharmacological agents that act on FGF dependent signaling pathways result in the stimulation of regeneration or prevent neurodegeneration in the PNS or CNS in a clinically significant way? Answering these and related questions will certainly provide a fruitful opportunity for many neuroscientists to substantively increase our appreciation of how a functioning nervous system is formed, maintained, and repaired in the organism.

References

Abraham JA, Mergia A, Whang JB, Tumpolo A, Friedman J, Hjerrild KA, Gospodarowicz D, Fiddes JC (1986) Nucleotide sequence of a bovine clone encoding the angiogenic protein basic fibroblast growth factor. Science 233: 545–548

Adashi EY, Resnick CE, Croft CS, May JV, Gospodarowicz D (1988) Basic fibroblast growth factor as a regulator of ovarian granulosa cell differentiation: a novel non-mitogenic role. Mol Cell Endocrinol 55: 14–17

Alterio J, Halley C, Brou C, Soussi T, Laurent M (1988) Characterization of a bovine acidic FGF cDNA clone and its expression in brain and retina. FEBS Lett 242: 41–46

Anderson KJ, Dam D, Lee S, Cotman CW (1988) Basic fibroblast growth factor prevents death of lesioned cholinergic neurons in vivo. Nature 332: 360–361

Baird A, Hsueh AJW (1986) Fibroblast growth factor as an intraovarian hormone: differential regulation of steriodogenesis by an angiogenic factor. Regulatory Peptides 16: 243–250

Baird A, Ling N (1987) Fibroblast growth factors are present in the extracellular matrix produced by endothelial cells in vitro: implications for a role of heparinase-like enzymes in the neovascular response. Biochem Biophys Res Commun 142: 428–435

Baird A, Schubert D, Ling N, Guillemin R (1988) Receptor- and heparin-binding domains of basic fibroblast growth factor. Proc Natl Acad Sci USA 85: 2324–2328

Biswas SB, Hammond RW, Anderson LD (1988) Fibroblast growth factors from bovine pituitary and human placenta and their functions in the maturation of porcine granulosa cells in vitro. Endocrinology 123: 559–566

Bouche G, Gas N, Pratish, Baldin V, Tauber JP, Teissie J, Amalric F (1987) Basic fibroblast growth factor enters the nucleolus and stimulates the transcription of ribosomal genes in ABAE cells undergoing G0 to G1 transition. Proc Natl Acad Sci USA 84: 6770–6774

Bovi PD, Basilico C (1987) Isolation of a rearranged human transforming gene following transfection of Kaposi sarcoma DNA. Proc Natl Acad Sci USA 84: 5660–5664

Bovi PD, Curatola AM, Kern FG, Greco A, Ittmann M, Basilico C (1987) An oncogene isolated by transfection of Kaposi's sarcoma DNA encodes a growth factor that is a member of the FGF family. Cell 50: 729–737

Bovi PD, Curatola AM, Newman KM, Sato Y, Moscatelli D, Hewick RM, Rifkin DB, Basilico C (1988) Processing, secretion, and biological properties of a novel growth factor of the fibroblast growth factor family with oncogenic potential. Mol Cell Biol 8: 2933–2941

Canalis E, Centrella M, McCarthy T (1988) Effects of basic fibroblast growth factor on bone formation in vitro. J Clin Invest 81: 1572–1577

Cho Ko, Skarnes WC, Minsk B, Palmieri S, Jackson-Grusby L, Wagner JA (1989) Nerve growth factor regulates gene expression by several distinct mechanisms. Cell Biol 9: 135–143

Claude, Parada IM, Gordon KA, D'Amore PA, Wagner JA (1988) Acidic fibroblast growth factor stimulates adrenal chromaffin cells to proliferate and to extend neurites but is not a long-term survival factor. Neuron 1: 783–790

Cordeiro PG, Seckel BR, Lipton SL, D'Amore PA, Wagner JA, Madison R (1989) Acidic fibroblast growth factor enhances peripheral nerve regeneration in vitro. Plast Reconstr Surg 83: 1013–1019

Coughlin SR, Barr PJ, Cousens LS, Williams LT (1988) Acidic and basic fibroblast growth factors stimulate tyrosine kinase activity in vitro. J Biol Chem 263: 988–993

Courty J, Danchel MC, Mereau A, Badet J (1988) Presence of basic fibroblast growth factor receptor in bovine brain membranes. J Biol Chem 263(3): 11217–11220

Damon DH, D'Amore PA, Wagner JA (1988) Sulfated glycosaminoglycans modify growth factor-induced neurite outgrowth in PC12 cells. J Cell Physiol 135: 293–300

Damon DH, Lobb RR, D'Amore A, Wagner JA (1989) Heparin potentiates the action of acidic fibroblast growth factor by prolonging its biological half-life. J Cell Physiol 138: 221–226

De Vitry F, Delaunoy JP, Tibault J, Lamande N, Legualt L, Selasalle A, Dopouey P (1983) Induction of oligodendrocyte-like properties in a primative hypothalamic cell line by cholesterol; an eye derived growth factor and brain extract. EMBO J 2: 199–203

Eccleston PA, Gunton DJ, Silberberg DH (1985) Requirements for brain cell attachment, survival and growth in serum-free medium: effects of extracellular matrix epidermal growth factor and fibroblast growth factor. Dev Neurosci 7: 308–322

Edgar D, Timpl R, Thoenen H (1984) The heparin-binding domain of laminin is responsible for its effects on neurite outgrowth and neuronal survival. EMBO J 3: 1463–1468

Feige JJ, Baird A (1988) Glycosylation of the basic fibroblast growth factor receptor. J Biol Chem 263: 14023–14029

Ferrara N, Schweigerer L, Neufeld G, Mitchell R, Gospodarowicz D (1987) Pituitary follicular cells produce basic fibroblast growth factor. Proc Natl Acad Sci USA 84: 5773–5777

Finklestein SP, Apostolides PJ, Caday CG, Prosser J, Philips MF, Klagsbrun M (1988) Increased basic fibroblast growth factor (bFGF) immunoreactivity at the site of focal brain wounds. Brain Res 20: 253–259

Folkman J, Klagsbrun M (1987) Angiogenic factors. Science 235: 442–447
Folkman J, Klagsbrun M, Sasse J, Wadzinski M, Ingber D, Vlodavsky I (1988) A heparin-binding angiogenic protein-basic fibroblast growth factor-is stored within basement membrane. Am J Pathol 130: 393–400
Friesel R, Maciag T (1988) Internalization and degradation of heparin binding growth factor-1 by endothelial cells. Biochem Biophys Res Commun 151: 957–964
Friesel R, Burgess WH, Mehlman T, Maciag T (1986) The characterization of the receptor for endothelial cell growth factor by covalent ligand attachment. J Biol Chem 261(17) 7581–7584
Gensburger C, Labourdett G, Sensenbrenner M (1987) Brain fibroblast growth factor stimulates the proliferation of rat neuronal precursor cells in vitro. FEBS Left 217: 1–5
Gimenez-Gallego G, Conn G, Hatcher VB, Thomas KA (1986) Human brain-derived acidic and basic fibroblast growth factors: amino terminal sequences and specific mitogenic activities. Biochem Biophys Res Commun 135: 541–548
Globus RK, Patterson-Buckendahl P, Gospodarowicz D (1988) Regulation of bovine bone cell proliferation by fibroblast growth factor and transforming growth factor beta. Endocrinology 123: 105–198
Gospodarowicz D (1974) Localization of a fibroblast growth factor and its effect alone and with hydrocortisone on 3T3 cell growth. Nature 249: 123–127
Gospodarowicz D, Cheng J (1986) Heparin protects basic and acidic FGF from inactivation. J Cell Physiol 128: 475–484
Gospodarowicz D, Neufeld G, Schweigerer L (1986) Fibroblast growth factor. Mol Cell Endocrinol 46: 187–205
Gospodarowicz D, Neufeld G, Schweigerer L (1987) Fibroblast growth factor: structural and biological properties. J Cell Physiol [Suppl] 5: 15–26
Greenberg ME, Greene LA, Ziff EB (1985) Nerve growth factor and epidermal growth factor induce rapid transient changes in proto-oncogene transcription in PC12 cells. J Biol Chem 260: 14101–14110
Halaban R, Kwon BS, Ghosh S, Bovis PD, Baird A (1988a) bFGF as an autocrine factor for human melanomas. Oncogene Res 3: 177–186
Halaban R, Langdon R, Birchall N, Cuono C, Baird A, Scott G, Moellmann G, McGuire J (1988b) Basic fibroblast growth factor from human keratinocytes is a natural mitogen for melanocytes. J Cell Biol 107: 1611–1619
Halperin J, Lobb R (1987) Effect of heparin-binding growth factors on monovalent cation transport in Balb/c 3T3 cells. Biochem Biophys Res Commun 144: 115–122
Hashimoto S (1988) K-252a, a potent protein kinase inhibitor blocks nerve growth factor-induced neurite outgrowth and changes in the phosphorylation of proteins in PC12h cells. J Cell Biol 107: 1531–1539
Hatten ME, Lynch M, Rydel RE, Sanchez J, Joseph-Silverstein J, Moscatelli D, Rifkin DB (1988) In vitro neurite extension by granule neurons is dependent upon astroglial-derived fibroblast growth factor. Dev Biol 125: 280–289
Hausch Ka CH, Linkhart TA, Olwin BB, Hausch Ka SD (1987) Growth factor control of skeletal muscle differentiation: commitment to terminal differentiation occurs in G1 phase and is repressed by fibroblast growth factor. J Cell Biol 105: 949–956
Heldin CH, Backstrom G, Ostman A, Hammacher A, Ronnstrand L, Rubin K, Nister M, Westermark B (1988) Binding of different dimeric forms of PDGF to human fibroblasts: evidence for two separate receptor types. EMBO J 7: 1387–1393
Hicks D, Courtois Y (1988) Acidic fibroblast growth factor stimulates opsin levels in retinal photoreceptor cells in vitro. FEBS Lett 234: 475–479
Huang SS, Huang JS (1986) Association of bovine brain-derived growth factor receptor with protein tyrosine kinase activity. J Biol Chem 261: 9568–9571
Huebner K, Ferrari AC, Bovi PB, Croce CM, Basilico C (1988) The FGF-related oncogene *K-FGF* maps to human chromosome region 11q13 possibly near *int*-2. Oncogene Res 3: 263–270
Imamura T, Tokita Y, Mitsui Y (1988) Purification of basic FGF receptors from rat brain. Biochem Biophys Res Commun 155: 583–590
Jakobovits A, Shackleford G, Varmus HE, Martin GR (1986) Two proto-oncogenes implicated in mammary carcinogenesis, *int-1* and *int-2*, are independently regulated during mouse development. Proc Natl Acad Sci USA 83: 7806–7810
Janet T, Miehe M, Pettmann B, Labourdette G, Sensenbrenner M (1987) Ultrastructural localization of fibroblast growth factor in neurons of rat brain. Neurosci Lett 80: 153–157

Jaye M, Howk R, Burgess W, Ricca GA, Chiu IM, Ravera MW, O'Brien SJ, Modi WS, Maciag T, Drohan WN (1986) Human endothelial cell growth factor: cloning, nucleotide sequence, and chromosome localization. Science 233: 541–545

Jaye M, Lyall RM, Mudd R, Schlessinger J, Sarver N (1988) Expression of acidic fibroblast growth factor cDNA confers growth advantage and tumorigenesis to Swiss 3T3 cells. EMBO J 7: 963–969

Jeanny JC, Fayein N, Moenner M, Chevallier B, Bqrritault D, Courtois Y (1987) Specific fixation of bovine brain and retinal acidic and basic fibroblast growth factors to mouse embryonic eye basement membranes. Exp Cell Res 171: 63–75

Kaibuchi K, Tsuda T, Kikuchi A, Tanimoto T, Yamashita T, Takai Y (1986) Possible involvement of protein kinase C and calcium ion in growth factor-induced expression of c-*myc* oncogene in Swiss 3T3 fibroblasts. J Biol Chem 261: 1187–1192

Kan M, DeSorbo D, Hou J, Hoshi H, Mansson PF, McKeehan WL (1988) High and low affinity binding of heparin binding growth factor to a 130-kDa receptor correlates with stimulation and inhibition of growth of a differentiated human hepatoma cell. J Biol Chem 263: 11206–11313

Kardami E, Spector D, Strohman RC (1988) Heparin inhibits skeletal muscle growth in vitro. Dev Biol 126: 19–28

Kimelman D, Kirschner M (1987) Synergistic induction of mesoderm by FGF and TGF-beta and the identification of an mRNA coding for FGF in the early *Xenopus* embryo. Cell 51: 869–877

Kimelman D, Abraham JA, Haaparanta T, Palisi TM, Kirschner MW (1988) The presence of fibroblast growth factor in the frog egg: its role as a natural mesoderm inducer. Science 242: 1053–1056

Klagsbrun M, Smith S, Sullivan R, Shing Y, Davidson S, Smith JA, Sasse J (1987) Multiple forms of basic fibroblast growth factor: amino-terminal cleavages by tumor cell and brain-derived acid proteinases. Proc Natl Acad Sci USA 84: 1830–1843

Kniss DA, Burry RW (1988) Serum and fibroblast growth factor stimulate quiescent astrocytes to re-enter the cell cycle. Brain Res 439(2): 281–288

Koizumi S, Contreras ML, Matsuda Y, Hama T, Lazarovici P, Guroff G (1988) K-252a: a specific inhibitor of the action of nerve growth factor on P12 Cells. Neuroscience 8: 715–721

Lander AD, Fujii DK, Gospodarowicz D, Reichardt LF (1982) Characterization of a factor that promotes neurite outgrowth: evidence linking activity to a heparin sulfate proteoglycan. J Cell Biol 94: 574–585

Leonard D, Ziff E, Greene L (1987) Identification and characterization of mRNAs regulated by nerve growth factor in PC12 cells. Mol Cell Biol 7: 3156–3167

Lipton SA, Wagner JW, Madison RD, D'Amore PA (1988) Acidic fibroblast growth factor enhances regeneration of processes by postnatal mammalian retinal ganglion cells in culture. Proc Natl Acad Sci USA 85: 2388–2392

Lobb RR (1988) Thrombin inactivates acidic fibroblast growth factor but not basic fibroblast growth factor. Biochemistry 27: 2572–2578

Lobb RR, Harper JW, Fett JW (1986) Purification of heparin-binding growth factors. Anal Biochem 154: 1–14

Magnaldo I, L'Allemain G, Chambard JC, Moenner M, Barritault D, Pouyssegur J (1986) The mitogenic signaling pathway of fibroblast growth factor is not mediated through polyphosphoinositide hydrolysis and protein kinase C activation in hamster fibroblasts. J Biol Chem 261: 16916–16922

Mergia A, Eddy R, Abraham JA, Fiddes JC, Shows TB (1986) The genes for basic and acidic fibroblast growth factors are on different human chromosomes. Biochem Biophys Res Commun 138: 644–651

Mioh H, Chen J (1987) Acidic heparin binding growth factor transiently activates adenylate cyclase activity in human adult arterial smooth muscle cells. Biochem Biophys Res Commun 146: 771–776

Moenner M, Chevallier B, Badet J, Barritault D (1986) Evidence and characterization of the receptor to eye-derived growth factor I: the retinal form of basic fibroblast growth factor on bovine epithelial lens cells. Proc Natl Acad Sci USA 83: 5024–5028

Mormede P, Baird A (1988) Estrogens, cyclic adenosine 3′5′-monophosphate, and phorbol esters modulate the prolactin response of GH3 cells to basic fibroblast growth factor. Endocrinology 122: 2265–2271

Morrison RS, De Vellis J (1981) Growth of purified astrocytes in a chemically defined medium. Proc Natl Acad Sci USA 78: 7205–7209

Morrison RS, De Vellis J, Lee YL, Bradshaw RA, Eng LF (1985) Hormones and growth factors induce the synthesis of glial fibrillary acidic protein in rat brain astrocytes. Neurosci Res 14: 167–176
Morrison RS, Sharma A, De Vellis J, Bradshaw RA (1986) Basic fibroblast growth factor supports the survival of cerebral cortical neurons in primary culture. Proc Natl Acad Sci USA 83: 7537–7541
Morrison RS, Keating RF, Moskal JR (1988) Basic fibroblast growth factor and epidermal growth factor exert differential trophic effects on CNS neurons. Neurosci Res 21: 71–79
Moscatelli D (1987) High and low affinity binding sites for basic fibroblast growth factor on cultured cells: absence of a role for low affinity binding in the stimulation of plasminogen activator. J Cell Physiol 131: 123–130
Moscatelli D (1988) Metabolism of receptor-bound and matrix-bound basic fibroblast growth factor by bovine capillary endothelial cells. J Cell Biol 107: 753–759
Moscatelli D, Presta M, Joseph-Silverstein J, Rifkin DB (1986) Both normal and tumor cells produce basic fibroblast growth factor. J Cell Physiol 129: 273–276
Moscatelli D, Joseph-Silverstein J, Manejias R, Rifkin DB (1987) *Mr* 25 000 heparin-binding protein from guinea pig brain is a high molecular weight form of basic fibroblast growth factor. Proc Natl Acad Sci USA 84: 5778–5782
Murphy PR, Sato R, Sato Y, Friesen HG (1988a) Fibroblast growth factor messenger ribonucleic acid expression in a human astrocytoma cell line: regulation by serum and cell density. Mol Endocrinol 2: 591–598
Murphy PR, Sato Y, Sato R, Friesen HG (1988b) Regulation of multiple basic fibroblast growth factor messenger ribonucleic acid transcripts by protein kinase C activators. Mol Endocrinol 2: 1196–1201
Navre M, Ringold GM (1988) A growth factor-repressible gene associated with kinase C-mediated inhibition of adipocyte differentiation. J Cell Biol 107: 279–286
Neufeld G, Gospodarowicz D (1985) The identification and partial characterization of the fibroblast growth factor receptor of basy hamster kidney cells. J Biol Chem 260: 13860–13868
Neufeld G, Gospodarowicz D (1986) Basic and acidic fibroblast growth factors interact with the same cell surface receptors. J Biol Chem 261: 5631–5637
Neufeld G, Gospodarowicz D (1988) Identification of the fibroblast growth factor receptor in human vascular endothelial cells. J Cell Physiol (Suppl) 136: 537–542
Neufeld G, Gospodarowicz D, Dodge L, Fujii DK (1987) Heparin modulation of the neurotrophic effects of acidic and basic fibroblast growth factors and nerve growth factor on PC12 cells. J Cell Physiol 131: 131–140
Neufeld G, Mitchell R, Ponte P, Gospodarowicz D (1988) Expression of human basic fibroblast growth factor cDNA in baby hamster kidney-derived cells results in autonomous cell growth. J Cell Biol 106: 1385–1394
Nieto-Sampedro M, Lim R, Hicklin DJ, Cotman CW (1988) Early release of glial maturation factor and acidic fibroblast growth factor after brain injury. Neuroscience 86: 361–365
O'Keefe EJ, Chiu ML, Payne RE (1988) Stimulation of growth of keratinocytes by basic fibroblast growth factor. J Invest Dermatol 90: 767–769
Otto D, Unsicker K, Grothe C (1987) Pharmacological effects of nerve growth factor and fibroblast growth factor applied to the transected sciatic nerve on neuron death in adult rat dorsal root ganglia. Neurosci Lett 83: 156–160
Olwin B, Hauschka SD (1988) Cell surface fibroblast growth factor and epidermal growth factor receptors are permanently lost during skeletal muscle terminal differentiation in culture. J Biol Chem 107: 761–769
Pelech SL, Olwin BB, Krebs EG (1986) Fibroblast growth factor treatment of Swiss 3T3 cells activates a subunit S6 kinase that phosphorylates a synthetic peptide substrate. Proc Natl Acad Sci USA 83: 5968–5972
Perraud F, Labourdette G, Miehe M, Loret C, Sensenbrenner M (1988) Comparison of the morphological effects of acidic and basic fibroblast growth factors on rat astroblasts in culture. Neurosci Res 20: 1–11
Pettmann B, Weibel M, Sensenbrenner M, Labourdette G (1985) Purification of two astroglial growth factors from bovine brain. FEBS Lett 189: 102–108
Pettmann B, Labourdette G, Weibel M, Sensenbrenner M (1986) The brain fibroblast growth factor (FGF) is localized in neurons. Neurosci Lett 68: 175–180
Prats H, Kaghad M, Prats AC, Klagsbrun M, Lelias JM, Liauzum, Chalon P, Tauber JP, Amalric F, Smith J (1989) High molecular mass forms of basic fibroblast growth factor are initiated by alternative CUG codons. Proc Natl Acad Sci USA 86: 1836–1840

Raeside JI, Berthelon MC, Sanchez P, Saez JM (1988) Stimulation of aromatase activity in immature porcine Leydig cells by fibroblast growth factor (FGF). Biochem Biophys Res Commun 151: 163–169

Ristow HJ, Messmer TO (1988) Basic fibroblast growth factor and insulin-like growth factor I are strong mitogens for cultured mouse keratinocytes. J Cell Physiol 137: 277–284

Rodan SB, Wesolowski G, Thomas K, Rodan GA (1987) Growth stimulation of rat Calvaria osteoblastic cells by acidic fibroblast growth factor. Endocrinology 121: 1917–1923

Rogelj S, Weinberg RA, Fanning P, Klagsbrun M (1988) Basic fibroblast growth factor fused to a signal peptide transforms cells. Nature 331: 173–176

Rogister B, Leprince P, Pettmann B, Labourdetter G, Sensenbrenner M, Moonen G (1988) Brain basic fibroblast growth factor stimulates the release of plasminogen activators by newborn rat cultured astroglial cells. Neurosci Lett 91: 321–326

Rosengart TK, Johnson WV, Friesel R, Clark R, Maciag T (1988) Heparin protects heparin-binding growth factor-I from proteolytic inactivation in vitro. Biochem Biophys Res Commun 152: 432–440

Rybal SM, Lobb RR, Fett JW (1988) Comparison of the effects of class 1 and class 2 heparin-binding growth factors on protein synthesis and actin mRNA expression in Balb/c3T3 cells. J Cell Physiol 136: 312–318

Rydel RE, Greene LA (1987) Acidic and basic fibroblast growth factors promote stable neurite outgrowth and neural differentiation of PC12 cells. J Neurosci 7: 3639–3653

Sakaguchi M, Kajio T, Kawahara K, Kato K (1988) Antibodies against basic fibroblast growth factor inhibit the autocrine growth of pulmonary artery endothelial cells. FEBS Left 233: 163–166

Saksela O, Moscatelli D, Sommer A, Rifkin DB (1988) Endothelial cell-derived heparan sulfate binds basic fibroblast growth factor and protects it from proteolytic degradation. J Cell Biol 107: 743–751

Sasada R, Kurokawa T, Iwane M, Igarashi K (1988) Transformation of mouse Balb/c 3T3 cells with human basic fibroblast growth factor cDNA. Mol Cell Biol 8: 588–594

Schreiber AB, Kenney J, Kowalski WJ, Friesel R. Mehlman T, Maciag T (1985) Interaction of endothelial cell growth factor with heparin: characterization by receptor and antibody recognition. Proc Natl Acad Sci USA 82: 6138–6142

Schubert D, Ling N, Baird A (1987) Multiple influences of a heparin-binding growth factor on neuronal development. J Cell Biol 104: 635–643

Schweigerer L, Malerstein B, Neufeld G, Gospodarpwicz D (1987) Basic fibroblast growth factor is synthesized in cultured retinal pigment epithelial cells. Biochem Biophys Res Commun 143: 934–940

Seed J, Hauschka SD (1988) Clonal analysis of vertebrate myogenesis. Dev Biol 128: 40–49

Shimsaki S, Emoto N, Koba A, Mercado M, Shibata F, Cooksey K, Baird A, Ling N (1988) Complementary DNA cloning and sequencing of rat ovain basic fibroblast growth factor and tissue distribution of its mRNA. Biochem Biophys Res Commun 157: 256–263

Shing Y, Folkman J, Sullivan R, Butterfield C, Murray J, Klagsbrun M (1984) Heparin affinity: purification of a tumor-derived capillary endothelial cell growth factor. Science 223: 1296–1298

Slack JMW, Darlington BG, Heath JK, Godsave SF (1987) Mesoderm induction in early *Xenopus* embryos by heparin-binding growth factors. Nature 326: 197–200

Smith R, Peters G, Dickson C (1988) Multiple RNAs expressed from the *int-2* gene in mouse embryonal carcinoma cell lines encode a protein with homology to fibroblast growth factors. EMBO J 7: 1013–1022

Spitz G, Roman D, Strauss A, Olson EN (1986) Serum and fibroblast growth factor inhibit myogenic differentiation through a mechanism dependent on protein synthesis and independent of cell proliferation. J Biol Chem 261: 9483–9488

Stemple DL, Mahanthappa NK, Anderson DJ (1988) Basic FGF induces neuronal differentiation, cell division, and NGF dependence in chromaffin cells: a sequence of envents in sympathetic development. Neuron 1: 517–525

Thomas PS (1980) Hybridization of denatured RNA and small DNA fragments transferred to nitrocellulose. Proc Natl Acad Sci USA 77: 5201–5205

Thomas KA (1988) Transforming potential of fibroblast growth factor genes. Trends Biochem Sci 13: 327–328

Thornton SC, Mueller SN, Levine EM (1983) Human endothelial cells: use of heparin in cloning and long-term serial cultivation. Science 222: 623–625

Togari A, Baker D, Dickens G, Guroff G (1983) The neurite-promoting effect of fibroblast growth factor on PC12 cells. Biochem Biophys Res Commun 114: 1189–1193

Togari A, Dickens G, Kuzuya H, Guroff G (1985) The effect of fibroblast growth factor on PC 12 cells. J Neurosci 5: 307–316

Tomaselli KJ, Damsky CH, Reichardt LF (1987) Interactions of a neuronal cell line (PC12) with laminin collagen IV and fibronectin: identification of integrin-related glycoproteins involved in attachment and process outgrowth. J Cell Biol 105: 2347–2358

Tsuda T, Kaibuchi K, Kawahara Y, Fukuzaki H, Takai Y (1985) Induction of protein kinase C activation and Ca^{2+} mobilization by fibroblast growth factor in Swiss 3T3 cells. FEBS Lett 3042 191: 205–210

Uhlrich S, Lagente O, Choay J, Courtois Y, Lenfant M (1986) Structure activity relationship of heparin: stimulation of non-vascular cells by a synthetic heparin pentasaccharide in cooperation with human acidic fibroblast growth factors. Biochem Biophys Res Commun 139: 728–732

Unsicker K, Reichert-Preibshch H, Schmidt R, Pettmann B, Labourdette G, Sensenbrenner M (1987) Astroglial and fibroblast growth factors have neurotrophic functions for cultured peripheral and central nervous system neurons. Proc Natl Acad Sci USA 84: 5459–5463

Vigny M, Ollier-Hartmann MP, Lavigne M, Fayein N, Neanny JC, Laurent M, Courtois Y (1988) Specific binding of basic fibroblast growth factor to basement membrane-like structures and to purified heparan sulfate proteoglycan of the EHS tumor. J Cell Physiol 137: 321–328

Vlodavsky I, Folkman J, Sullivan R, Fridman R, Ishai-Michaeli R, Sasse J, Klagsbrun M (1987) Endothelial cell-derived basic fibroblast growth factor: synthesis and deposition into subendothelial extracellular matrix. Proc Natl Acad Sci USA 84: 2292–2296

Wagner JA, D'Amore PA (1986) Neurite outgrowth induced by an endothelial cell mitogen isolated from retina. J Cell Biol 103: 1363–1367

Walicke PA (1988a) Basic and acidic fibroblast growth factors have trophic effects on neurons from multiple CNS regions. J Neurosci 8: 2618–2627

Walicke PA (1988b) Interactions between basic fibroblast growth factor (FGF) and glycosaminoglycans in promoting neurite outgrowth. Exp Neurol 102: 144–148

Walicke PA, Baird (1988) Neurotrophic effect of basic and acidic fibroblast growth factors are not mediated through glial cells. Dev Brain Res 40: 711–779

Walicke P, Cowan WM, Ueno N, Baird A, Guillemin R (1986) Fibroblast growth factor promotes survival of dissociated hippocampal neurons and enhances neurite extension. Proc Natl Acad Sci USA 83: 3012–3016

Walicke PA, Feige JJ, Baird A (1989) Characterization of the neuronal receptor for basic fibroblast growth factor and comparison to receptors on mesenchymal cells. J Biol Chem 264: 4120–4126

Westall FC, Lennon VA, Gospodarowicz D (1978) Brain-derived fibroblast growth factor: identity with fragment of the basic protein of myelin. Proc Natl Acad Sci USA 75: 4675–4678

Wilkinson DG, Peters G, Dickson C, McMahon AP (1988) Expression of the FGF-related protooncogene *int-2* during gastrulation and neurulation in the mouse. EMBO J 7: 691–695

Winkles JA, Friesel JA, Burgess WH, Howk R, Mehlman T, Weinstein R, Maciag T (1987) Human vascular smooth muscle cells both express and respond to heparin-binding growth factor 1 (endothelial cell growth factor). Proc Natl Acad Sci USA 84: 7124–7128

Wu DK, Maciac TA, De Vellis J (1988) Regulation of neuroblast proliferation by hormones and growth factors in chemically defined medium. J Cell Physiol 136: 367–372

Yoshida T, Miyagawa K, Odagiri H, Sakamoto H, Little PFR, Terada M, Sugimura T (1987) Genomic sequence of *hst*, a transforming gene encoding a protein homologous to fibroblast growth factors and the *int-2* encoded protein. Proc Natl Acad Sci USA 84: 7305–7309

Zhan X, Bates B, Hu X, Goldfarb M (1988) The human *FGF-5* oncogene encodes a novel protein related to fibroblast growth factors. Mol Cell Biol 8: 3487–3495

Dissecting the Mode of Action of a Neuronal Growth Factor

S. Halegoua, R. C. Armstrong, and N. E. Kremer

1 Introduction

The neuronal growth factors nerve growth-factor (NGF) and fibroblast growth factor (FGF) play important roles in the development of both the peripheral and central nervous systems. These agents are survival factors in vivo and in neuronal cell culture. They appear to promote the differentiation of different classes of neurons and may also play some part in the guidance of neuronal process outgrowth. It is apparent that neuronal growth factors, as exemplified by NGF and FGF, encompass an enormous range of actions, making elucidation of the underlying molecular mechanisms a formidable task. However, a multidisciplinary approach has been brought to bear on this task, and studies of neuronal

Department of Neurobiology and Behavior, State University of New York at Stony Brook, Stony Brook, NY 11794-5230, USA

Current Topics in Microbiology and Immunology, Vol. 165

growth factor action have gained from as well as contributed insights to other fields.

Most of our present understanding of the mechanisms underlying NGF and FGF actions comes from cell culture studies where the environment can be tightly controlled. Two types of culture systems have predominated in these studies, primary neuronal cell culture and transformed cell lines. Primary cultures have provided a rich description of the phenotypic changes that occur in response to neuronal growth factor treatment. This culture system is also particulary well-suited to the study of neuronal growth factor effects on survival. Ironically, molecular studies of NGF and FGF action using primary neuronal cultures have been hampered by the very same property that made these cultures appealing in the first place. That is, because primary cultures require NGF for survival, it is difficult if not impossible to study the effects on survival separate from those involved in differentiation. An additional limitation of the primary culture systems lies in their previous exposure to neuronal growth factors in vivo, thus eliminating the ability to study initial effects of the growth factors on the naive cells. The establishment of the PC12 clonal cell line by GREENE and TISCHLER in 1976 circumvented these problems and provided an alternative, manipulatable culture system for molecular studies of NGF and FGF action. With these cells, neuronal growth factors are necessary and sufficient for neuronal differentiation, but the undifferentiated cells grow continuously in culture in the absence of these factors. This latter property is particularly useful from the point of view of introducing foreign DNA into NGF responsive cells in order to manipulate the system in specific ways and to study the growth factor effects on gene expression.

As in primary cell cultures, the changes in PC12 cell physiology after treatment with NGF and FGF are both dramatic and complex. Unraveling the puzzle of the mechanism of such changes has been a long and exciting, although at times perplexing, endeavor involving a unique signal transduction process. This chapter will focus primarily on revelations from studies of NGF and FGF actions in PC12 cells. Events apparently underlying neuronal growth factor actions on PC12 cells will be extrapolated where appropriate to primary neuronal cultures or in vivo systems.

2 PC12 Cell Culture System

The PC12 cell line, originally derived from a spontaneous rat pheochromocytoma (GREENE and TISCHLER 1976), displays basic phenotypic traits associated with both adrenal chromaffin cell and sympathetic neuron characteristics (for review see GREENE and TISCHLER 1982). In the following sections, we will concentrate on the properties of PC12 cells pertinent to their use in studies of NGF and FGF action.

In the absence of NGF or FGF, PC12 cells grow indefinitely in cell culture. When treated with NGF, PC12 cells undergo a dramatic transformation strikingly along the lines of differentiating sympathetic neurons. For example, PC12 cells undergo morphologic differentiation involving the formation of long neuronal processes (neurites) resembling axons, including the presence of prototypic growth cones. The neurites appear after at least a 2 day delay and elongate in the continual presence of NGF. Concurrent with the formation of neurites, PC12 cells acquire the ability to generate sodium based action potentials. In addition to sodium channels, a variety of other voltage dependent ion channels appear at the neuritic growth cone. Neurotransmitter synthesis increases after NGF treatment as evidenced by increases in the amounts of both catecholamines and acetylcholine, and these can be realeased in a vesicular fashion at the neuritic growth cone.

Most studies of the differentiation of PC12 cells have involved NGF. However, it has been suggested recently that FGF is also a neuronal growth factor for CNS neurons and can affect differentiation in the adrenosympathetic lineage (see WAGNER, this volume). Surprisingly, FGF also induces neuronal differentiation of PC12 cells, and this induction is in many respects similar to the differentiation by NGF. Importantly, not all growth factors cause the differentiation of PC12 cells. The inability of some growth factors (e.g. epidermal growth factor, EGF; insulin) to cause differentiation of PC12 cells can be used to advantage to identify those molecular events which are critical to achieving neuronal differentiation. Therefore, the effects of NGF and FGF have been compared to "non-differentiating" growth factors.

3 Signal Transduction

Differentiation induced by growth factors involves dramatic changes in major cellular processes. These changes are evident from as early as seconds to as long as days in the continual presence of the growth factor. Evidence has accumulated, as will be discussed below, for a model of growth factor action in which a variety of cellular control mechanisms are coordinately recruited to bear on the process of differentiation induced by NGF (Fig. 1). The binding of a neuronal growth factor to its receptor results in the stimulation of multiple second messenger pathways. Stimulation of these pathways causes a series of post-translational modifications of specific proteins, which lead to immediate changes in the cellular phenotype. Substrates for post-translational modification include DNA binding proteins which, when modified, elicit specific gene expression changes. The changes in gene expression have been compared to changes that occur during viral infection of cells. They are initially seen as a transient wave of very early gene inductions which have been termed "immediate early genes" and which, for

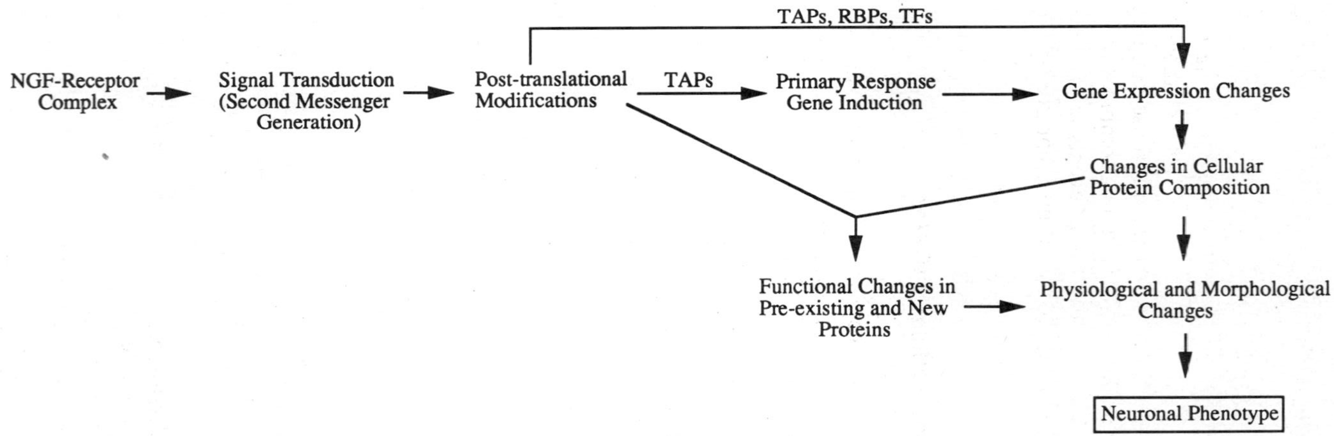

Fig. 1. Model for the molecular events underlying the NGF induced program for neuronal differentiation. *TAP*, transcription activating protein; *RBP*, RNA binding protein; *TF*, translation factor

mitogenic growth factor actions, have also been termed "primary response genes" (YAMAMOTO and ALBERTS 1976; LIM et al. 1987). The protein products of these genes have been generally found to be transcriptional trans-activators which in various combinations result in the control of the expression of a variety of genes, including their own. The initial induction of primary response genes leads to subsequent waves of gene inductions and repressions, which result in changes in the levels of proteins important to the differentiated phenotype. Other changes in gene expression caused by initial post-translational events include those resulting from changes in mRNA stability and/or translation. The ongoing combination of transcriptional, post-transcriptional, and post-translational events coordinate the differentiation process under the reversible and continual control of the neuronal growth factor.

3.1 NGF or NGF-Receptor as an Intracellular Effector

In some way, the binding of NGF to its receptor is transduced into an intracellular response encompassing a sequence of complex changes in the cell over a long time span. The question of whether NGF acts more directly to elicit these changes, such as through the internalization of the NGF-receptor complex, or acts indirectly through the generation of second messengers has been labored over for almost 20 years. It is certainly clear that both events occur; the NGF-receptor complex is internalized and various second messengers are generated in response to NGF treatment. However, the mechanism by which these events occur and the specific changes in cellular physiology brought about by these events have only begun to be dissected.

In 1974, HENDRY et al. provided evidence that NGF is taken up at sensory nerve terminals and retrogradely transported to the cell body. Recent findings have verified this result in vivo (JOHNSON et al. 1987). CAMPENOT (1977) has shown that NGF can act at nerve terminals to promote survival of the neuronal cell body in culture. These results have been taken to imply that NGF itself has an intracellular site of action. It is clear that NGF binds to cell surface receptors followed by vesicular internalization of the NGF-receptor complex. There have been two schools of thought as to the fate and role of this complex. In one view, the complex is internalized and transported toward the cell body. The complex may then be degraded in the lysosomes or the complex may be dissociated, followed by recycling of these components back to the cell surface. Ample evidence exists for both of these processes (see CHAO this volume) but, in this view, NGF-receptor complexes do not exert physiological effects after they are internalized. In the second view, either NGF or the NGF-receptor complex is transported to sites in the cell body where it may exert important biological actions.

If NGF has an intracellular site of action, it is likely to be in a complex with its membrane receptor, since injection of either free NGF or antibodies directed against NGF into PC12 cells does not have any apparent effects on NGF action. One possible site of NGF-receptor action which has received much attention is the nucleus. This site has been appealing since it could explain NGF's ability to elicit transcriptional control. Two types of studies have lent support to this hypothesis. In the first, BRADSHAW and colleagues (ANDRES et al. 1977) presented evidence for the existence of NGF receptors (in the form of ^{125}I-NGF binding sites) in the nucleus and, more specifically, associated with chromatin. YANKNER and SHOOTER (1979) subsequently presented evidence that the NGF-receptor complex became associated with the nucleus after internalization. This was seen by following both ^{125}I-NGF localization and NGF binding sites to the nuclear (Triton X-100 insoluble) fraction. However, a convincing study by BOTHWELL and colleagues (BOTHWELL et al. 1980) argued that these receptors were bound only to the cytoskeleton which was contaminating YANKNER and SHOOTER's nuclear preparations. Thus, the data to date do not appear to support the notion of NGF and/or its receptor having a direct nuclear effect.

A last possible action of the internalized NGF-receptor complex which has not yet been experimentally addressed is that of a second messenger generator. It must be kept in mind that the internalized complex would exist in an endocytosed, membrane-bound vesicle whose orientation is inside out. NGF would remain mostly bound to the receptor since it is trapped in the vesicle. This binding might then be continually transduced into a signal generated to the cell

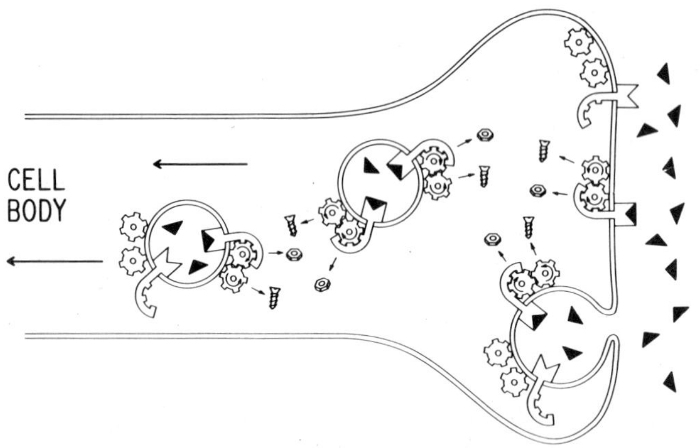

Fig. 2. A possible process for internalization and retrograde transport of functional NGF signal transduction machinery. Illustrated is a neurite terminal containing NGF (*filled triangles*) and its receptor (transmembrane molecule with extracellular ligand binding site and cytoplasmic region), as well as peripherally membrane-bound signal transducing molecules (represented as *gears*). Binding of NGF to receptor results in a conformational change in receptor such that it engages a signal transducing molecule, which may in turn engage another one or more signal transducers. Signal transduction results in the generation of various second messengers (represented as *nuts* and *bolts*) resulting from the activation of one or more signal transducers. Upon binding of NGF, the NGF-receptor complex is also endocytosed. The internalized vesicle containing NGF and the receptor binding domain inside, together with the signal transducers (facing the cytoplasm), is retrogradely transported to the cell body. During and after transport to the cell body, NGF binding to the receptor continues to result in stimulation of the signal transduction machinery and thus in the continual production of second messengers

interior (Fig. 2). In this way, NGF could exert influences similar to those from the cell surface, over the long distances of its intracellular transport (i.e., from the nerve terminal to the cell body). The experiments involving microinjections of NGF or anti-NGF mentioned above have not addressed this possibility because the injected proteins would not be expected to have entered the vesicular compartment containing NGF, its receptor, and the NGF-receptor complex.

3.2 Cyclic Adenosine Monophosphate

The search for second messengers of NGF action has resulted in reports on a variety of second messenger stimulations. The second messenger initially suggested to underline NGF action was cyclic adenosine 3′–5′ monophosphate (cAMP). Cyclic AMP was shown to rise transiently following NGF treatment of PC12 cells (Schubert and Whitlock 1977; Schubert et al. 1978) and sympathetic and sensory neurons (Nikodijevic et al. 1975; Narumi and Fujita 1978; Skaper et al. 1979). However, other investigators reported no such increases (Frazier et al. 1973; Hatanaka et al. 1978; Lakshmanan 1978). Derivatives of cAMP have also been reported to mimic aspects of NGF action. Increased choline acetyltransferase (Schubert et al. 1977) and ornithine decarboxylase (Hatanaka et al. 1978) activities, increased cell-substratum adhesion (Schubert et al. 1978), specific alterations in cell surface lectin binding properties and stimulation of neurite outgrowth (Schubert et al. 1978), stimulation of Ca^{++} efflux (Schubert et al. 1978), and modulation of both protein synthesis (Garrels and Schubert 1979) and protein phosphorylation (Halegoua and Patrick 1980; Yu et al. 1980; and see also below) all occur in response to cAMP treatment. These actions suggested that differentiation by NGF was in part or in whole carried out through cAMP and led to the proposal of a specific model of NGF action (Schubert et al. 1978). The model suggests that NGF induces an increase in the intracellular level of cAMP, causing an increase in Ca^{++} mobilization, which in turn produces a structural change in the cell membrane. This membrane alteration enhances cell-substratum adhesion and leads to neurite extension in PC12 cells.

Several groups further analyzed and compared the cAMP and NGF responses. With respect to neurite outgrowth, although intracellular cAMP elevation generated short-term neurites and potentiated the induction of neurites by NGF, neither continued extension of these processes nor their potentiation proceeded beyond 3 days. In contrast to NGF, cAMP-induced neurites were not dependent upon RNA synthesis, and cAMP treatment was not able to induce "priming" of PC12 cells to NGF-stimulated neurite regeneration (Gunning et al. 1981; also see Greene and Shooter 1980 for review). Analogues of cAMP which competitively inhibit cAMP-mediated activation of the cAMP dependent protein kinase (A-kinase) were able to block cAMP actions on PC12 cells but not NGF-induced neurite outgrowth (Richter-Landsberg and Jastorff 1986). Analyses of gene expression changes stimulated by NGF have demonstrated that some are not affected by cAMP alone (for example see Cho et al. 1989; Leonard et al. 1987). These results have argued against the involvement of cAMP in NGF action. However, a detailed analysis of tyrosine hydroxylase (TH) and other protein phosphorylations (see Sect. 3.4), indicate that NGF stimulates a subset of phosphorylations through the action of A-kinase. Taken together, the available data suggest that cAMP plays a role in some but not all NGF actions and that cAMP may act in concert with other second messengers (see below) to carry out NGF action. In an accessory role, cAMP may not appear to be either necessary or sufficient for some NGF actions, for example, long-term neurite outgrowth or the activation of TH leading to enhanced catecholamine synthesis. That cAMP plays an essential role in at least one prominent NGF action is indicated by the finding that the inhibitor protein of A-kinase blocks the induction of Na^+ channel activity and thus the development of an action potential mechanism in response to NGF (Kalman et al. 1990).

3.3 Phospholipid Metabolism

The turnover of phosphatidylinositol (PI) is another second messenger pathway suggested to play a role in NGF actions. In this pathway, the inositol lipids are broken down to liberate at least two second messengers, diacylglycerol which is an activator of protein kinase C (C-kinase), and inositol trisphosphate (IP_3) which causes a rise of intracellular free Ca^{++} liberated from intracellular stores (for review see Abdel-Latif 1986). Lakshmannan (1979) first reported increased labeling of the phosphoinositides in rat superior cervical ganglion neurons after 4 h of exposure to NGF. This finding was subsequently reproduced and extended in PC12 cells. $^{32}PO_4$ labeling (Traynor et al. 1982) or [^{3}H]myoinositol labeling (Traynor 1984) of each of the PIs (PI, PIP, PIP_2) in PC12 cells was specifically increased relative to that of phosphatidylcholine or phosphatidylethanolamine within 10 min of NGF treatment, a finding which is indicative of increased PI turnover. Similar effects were seen with other agents that stimulated either cell-substrate adhesion or neurite outgrowth in PC12 cells including agents which raise cAMP (Traynor 1984). These data were suggested to lend support to the model for NGF action proposed by Schubert et al. (1978) discussed above. The model was thus extended to identify PI turnover as either the link between cAMP and Ca^{++} mobilization or as a step subsequent to cAMP-mediated Ca^{++} mobilization previously suggested to be caused by NGF.

More recently, investigators have examined the breakdown of the phosphoinositides more directly by examining the release of [^{3}H]inositol phosphates from cultured bovine chromaffin cells (Pandiella-Alonso et al. 1986) or from PC12 cells (Contreras and Guroff 1987) in which these lipids were prelabeled. In both studies, NGF increased the production of labeled IP, IP_2, and IP_3 after as little as 15 s (Contreras and Guroff 1987) of treatment. Studies in our laboratory (J. Cremins, R. Armstrong, J. Whitbread, T. Whitford and S. Halegoua, in preparation) have addressed the question of cAMP involvement in this NGF effect on PC12 cells. NGF, FGF, and cAMP were each found to stimulate increased specific $^{32}PO_4$ labeling of PI, PIP, PIP_2, and phosphatidic acid. NGF-stimulated PI turnover was found to consist of two components, a minor cAMP dependent component and a major cAMP independent one. The cAMP dependent PI turnover could be distinguished by its sensitivity to pertussis toxin and its requirement for A-kinase. These results provide further support for the suggestion that cAMP plays at least an accessory role in NGF action and helps to explain some of the differences between cAMP and NGF actions.

Several studies have suggested a role for Ca^{++} in the PI turnover response to NGF. NGF-stimulated PI turnover was dependent upon extracellular Ca^{++} (PANDIELLA-ALONSO et al. 1986; CONTRERAS and GUROFF 1987), and an increased influx of extracellular Ca^{++} also elicited PI turnover in PC12 cells (TRAYNOR 1984; CONTRERAS and GUROFF 1987). PANDIELLA-ALONSO et al. (1986) reported that NGF rapidly (within 30 s) caused a rise in intracellular Ca^{++} in both PC12 cells and cultured adrenal chromaffin cells, as seen using the fluorescent Ca^{++} indicators quin-2 (on a population of cells) and fura-2 (at the single cell level). The Ca^{++} mobilizing effect of NGF was unusual as compared to depolarizing agents or neurotransmitters in that it was small, had a slow rise time, and a delayed onset (30–40 s) after NGF addition (see also LAZAROVICI et al. 1989). NGF-stimulated Ca^{++} mobilization was also dependent upon extracellular Ca^{++}, however, it could proceed if Ca^{++} was removed during the lag phase and beyond. From these and other results, the authors suggested that NGF-induced changes in Ca^{++} consist of two phases, one which involves an influx of extracellular Ca^{++} through channels other than the voltage dependent Ca^{++} channel and the other which involves IP_3-mediated Ca^{++} release from intracellular stores. Long-term Li^{+} treatment of cells interferes with PI turnover in a variety of cell types. Such treatment reversibly blocks NGF-induced neurite outgrowth and NGF-stimulated microtubule protein phosphorylation (BURSTEIN et al. 1985). Interestingly, the Li^{+} block of neurite growth could be reversed by depolarization-mediated Ca^{++} entry (S. HALEGOUA, unpublished work). The potential involvement of PI turnover and/or Ca^{++} in the blocking action of Li^{+} remains to be determined.

One result of the breakdown of the PI lipids is the liberation of diacylglycerol consisting predominantly of 1-stearoyl-2-arachidonoylglycerol, which is then rapidly converted to phosphatidic acid. However, when the production of diacylglycerol from [^{3}H]myristate or from [^{3}H]arachidonate labeled cells was examined, NGF-induced production of [^{3}H]myristate labeled diacylglycerol was more easily seen (CHAN et al. 1989). This result suggested that NGF could stimulate the breakdown of phospholipids other than PI. In the same study, the turnover of glycosylphosphatidylinositol was examined in [^{3}H]glucosamine labeled cells. Increased turnover of this lipid as well as production of inositol glycan was detected in response to NGF. However, the relative proportion of the total NGF-stimulated diacylglycerol derived from this lipid was not determined. This result led to the suggestion that some NGF actions may be mediated through the breakdown of glycosylphosphatidylinositol (as suggested for other hormone effectors, see SALTIEL and CUATRECASAS 1988) with the resulting production of diacylglycerol followed by C-kinase activation.

An alternate mode of phospholipid metabolism which is exploited by the cell to generate a different range of second messengers is stimulation of phospholipase A2. The resulting release of arachidonic acid from some phospholipids is followed by this fatty acid's metabolism to various derivatives. The ability of NGF to stimulate arachidonic acid metabolism in PC12 cells was examined by DEGEORGE et al. (1988). It was found that NGF could stimulate the production of a variety of arachidonic acid derivatives either in homogenates or in cell culture, within as little as 5 min of treatment. Inhibitors of phospholipase A2 as well as of the lipoxygenase pathway for arachidonic acid metabolism were able to block NGF-induced neurite outgrowth. Inhibitors of the cyclooxygenase pathway for arachidonate metabolism had no such effect. Although the use of such pharmacological agents in long-term culture must be viewed with caution, the results suggest that the lipoxygenase pathway of lipid metabolism may be required for the mediation of neurite growth by NGF. Clearly further studies are warranted to elucidate the potential involvement of the lipoxygenase pathway in NGF action.

3.4 Protein Phosphorylation

A post-translational event which seems to underlie nearly all second messenger pathways is the phosphorylation of specific proteins. Resulting changes in these protein functions are a major means by which physiological changes are brought about. For the NGF system, studies of these processes have generally taken three approaches. One approach has focused on the identification of unknown proteins whose phosphorylation either in vivo or in cell extracts is enhanced by in vivo NGF treatment. Identification of these kinase substrates provides a basis for subsequent studies on determination of their roles in NGF action and on the identity of the protein kinase and second messenger system which controls the phosphorylation. A second approach has attempted to identify NGF-stimulated protein kinase activities assayed by using artificial or natural kinase substrates and measuring their phosphorylation, carried out in vitro or in situ. In a third approach, known proteins already identified as, or suspected of, playing a role in NGF action have been further analyzed for possible phosphorylation by NGF-stimulated protein kinases.

One study, using in vivo labeling of PC12 cells and of cultured sympathetic ganglia with $^{32}PO_4$, revealed increased labeling by up to 70% of a 30 kD nuclear protein by NGF treatment (YU et al. 1980). This protein was purified and determined to be a non-histone chromosomal protein, termed slow migrating protein (SMP). The protein phosphorylation, exclusively on serine residues, was seen also in response to EGF, TPA, dbcAMP, and FGF treatment (TOGARI et al. 1985). These effects were subsequently reproduced in cell free extracts of NGF treated cells (Nakanishi and Guroff 1985). The kinase responsible for this phosphorylation was determined not to be similar to C-kinase, Ca^{++}/calmodulin dependent protein kinase (CAMP kinase), or A-kinase. However, a cascade mechanism for SMP phosphorylation, involving A-kinase or C-kinase, remains a possibility.

A series of studies from our laboratory has examined protein phosphorylations carried out within minutes after NGF treatment of PC12 cells. Increased $^{32}PO_4$ labeling of a variety of nuclear and non-nuclear proteins was first seen as five radiolabeled bands by one dimensional SDS-gel electrophoretic analysis (Halegoua and Patrick 1980). These bands were identified as the known proteins, TH, ribosomal protein S6, histones H1a and H3, and the non-histone chromosomal high mobility group protein HMG 17. The turnover rate of the stimulated labeling was even higher than that of the basal labeling of these and many other proteins. This result implied that the increased protein phosphorylation resulted from a stimulated kinase and that the NGF-stimulated phosphorylations were relatively susceptible to phosphatase action. Stimulation of these protein phosphorylations was both rapid and prolonged, with maximal phosphorylation occurring within 15 min and remaining elevated for at least 3 days during continuous NGF treatment of the cells. In an effort to identify the protein kinase involved in this NGF effect, phosphorylation patterns generated by treatment with NGF, EGF, insulin, and elevation of intracellular cAMP were compared. NGF and cAMP resulted in the same pattern of protein phosphorylations, which was different from those of EGF and insulin. Results from these experiments as well as from additivity experiments using these agents to stimulate phosphorylation of ribosomal protein S6 led to the suggestion that NGF stimulated the phosphorylation of the A-kinase. From this study and extrapolating from work on other systems, it was suggested that some of these phosphorylations played roles in known NGF actions such as increased protein synthesis (for ribosomal protein S6) and changes in DNA structure and transcriptional activation (for the chromosomal proteins). A new action of NGF was also suggested, rapidly increased catecholamine synthesis resulting from the phosphorylation and activation of TH.

The activation of TH by NGF-stimulated phosphorylation was suggested from the results of two independent groups. In one case, the activity of TH was found to be elevated in extracts of cells which had been treated with NGF, in a time course similar to that of TH phosphorylation (Lee et al. 1985). A similar time course of TH activation was seen in PC12 cells in response to NGF (McTigue et al. 1985) using an assay for TH activity which measures the rate of dopa accumulation in living cells. Acute changes in TH activity have been seen both in vivo and in vitro in response to a variety of agents and culture conditions. These acute changes have in some cases been shown to be a direct consequence of phosphorylation through different second messenger systems (for review see Zigmond et al. 1989). In order to characterize the kinases responsible for TH phosphorylation, its multisite phosphorylation in PC12 cells has been studied in detail using phosphopeptide mapping. Four distinct sites of TH phosphorylation were observed (McTigue et al. 1985). Differential, site specific phosphorylation among these four sites was achieved by NGF, EGF, cAMP, depolarization, and phorbol esters, indicating that at least four kinases could phosphorylate TH. Each of these phosphorylations were at single serine residues (McTigue et al. 1985; J. Cremins and S. Halegoua, unpublished work). NGF stimulated the phosphorylation of two peptides (T1 and T3), in common with those stimulated by cAMP (T1 and T3) and phorbol esters (T3). A subsequent study concluded that these two sites were phosphorylated in response to stimulation of A-kinase (T1) and C-kinase (T3) (Cremins et al. 1986), because NGF-stimulated phosphorylation of T1 was specifically blocked in an A-kinase deficient PC12 mutant (A126–1B2) and that of T3 was specifically blocked by drugs which also blocked both phorbol ester and dioctanoyl glycerol-mediated phosphyorylation of T3. It is interesting that FGF stimulated the phosphorylation of only site T3 (D. Damon, J. Wagner, and S. Halegoua, unpublished work). The phosphorylation at site T1 is not stimulated in response to FGF, suggesting that the action of FGF may not involve the A-kinase. A-kinase-mediated phosphorylation of site T3 (McTigue et al. 1985) is believed to occur by indirect activation of C-kinase through a cAMP/PI turnover crosstalk pathway (discussed in Sect. 3.3), since this pathway is specifically blocked by pertussis toxin (J. Cremins et al., in preparation). TH is an excellent substrate in vitro for both the A-kinase and CAMP kinase II. The major sites of phosphorylation by these kinases are serine residues at position 40 and 19, respectively (Zigmond et al. 1989). These sites correspond to those in phosphopeptide T1 and T2 (see above), which are phosphorylated in response to cAMP and Ca^{++} elevations, respectively, in PC12 cells (J. Haycock, personal communication). Thus, it appears likely that A-kinase and CAMP kinase II directly phosphorylate T1 and T2 in vivo. In contrast, TH is poorly phosphorylated in vitro by C-kinase and the site of phosphorylation is the same as that of A-kinase (Albert et al. 1984). This is different from the site phosphorylated in vivo in response to phorbol esters and diacylglycerol derivatives (T3) (see above). Thus it is quite possible that C-kinase does not directly phosphorylate TH in vivo, but rather mediates its phosphorylation through a kinase cascade. The kinase capable of T3 phosphorylation in vitro has not yet been described.

In an alternative approach for identifying NGF-stimulated protein kinases, TH was used as a substrate for phosphorylation in PC12 cell extracts (Rowland et al. 1987). Treatment of cells with NGF for as little as 1–3 min resulted in the stimulation of a kinase activity which could phosphorylate serine residues of exogenously added TH in cell extracts. This soluble kinase activity, designated N-kinase, was partially purified and characterized, having an apparent molecular mass of 22–25 kDa. It could also phosphorylate histone H1 and ribosomal protein S6 in vitro. N-kinase activity stimulated by NGF was independent of cAMP and Ca^{++}, was resistant to inhibitors of C-kinase and calmodulin, and was not down regulated by prolonged phorbol myristate acetate (PMA) treatment (which down regulates the cellular level of its receptor, C-kinase). The N-kinase was thus suggested to be distinct from A-kinase, C-kinase, and

CAMP kinase. However, cAMP elevation, Ca^{++} elevation, and phorbol ester derivatives can all result in N-kinase activation in vivo, suggesting that this kinase may be activated in a cascade manner. N-kinase has been suggested to be the same as a previously described S6 kinase (Blenis and Erikson 1986) discussed below. Recently, N-kinase was found to be inhibited by 6-thioguanine (Volonte et al. 1989). The inability of these inhibitors to block NGF-stimulated TH phosphorylation in PC12 cells suggests that this kinase is not responsible for the TH phosphorylation seen in vivo. However, this major kinase activity is likely to have other as yet unidentified endogenous substrates which are important for mediating some NGF actions. This is indicated by, for example, the ability of 6-thioguanine to inhibit both the induction of ornithine decarboxylase (ODC) and neurite regeneration from PC12 cells (Volonte et al. 1989).

Investigators in two laboratories have carried out similar in vitro studies to determine the kinases responsible for ribosomal protein S6 phosphorylation in response to NGF. In one set of studies (Matsuda et al. 1986; Matsuda and Guroff 1987), PC12 cell extracts were assayed for kinase activity which could phosphorylate S6. S6 phosphorylation was found to be highly stimulated (10- to 20-fold) in extracts from NGF treated cells relative to control cells. The time course of NGF stimulation of S6 kinase was similar to that seen in vivo (see above), with complete activation seen within 15 min. The kinase, purified from PC12 cells, has a molecular mass of 45 kDa based on Sephadex G200 chromatography and is specific for S6 phosphorylation. It is sensitive to inhibition by glycerophosphate and EGTA and by NaF, but is not affected by inhibitors of A-kinase, C-kinase, or CAMP kinase II. Based on the above properties, it does not appear to be related to any previously characterized S6 kinase. That the kinase may be activated by phosphorylation is suggested by the shift of the kinase toward a more acidic pI concomitant with its activation by NGF. Alkaline phosphatase treatment restores both the original pI and activity level. Elevation of cAMP also results in the S6 kinase activation in a manner which is not additive with NGF, a result which is consistent with those obtained in vivo (see above). Since A-kinase can activate S6 kinase by in vitro phosphorylation, it was suggested that NGF may activate S6 kinase through a kinase cascade involving the A-kinase.

Another investigation has revealed an S6 kinase with different properties (Blenis and Erikson 1986). While the NGF activation of S6 kinase activity seen by Matsuda and Guroff (1987) was transient (dropping to baseline within 2 h), this second S6 kinase activity was stimulated during at least 10 h of continuous NGF treatment. Furthermore, the chromatographic properties and inhibition profiles of the two activities differed markedly. Ion exchange chromatography showed two peaks of activity stimulated by NGF. The major peak of activity had the same elution profile as the S6 kinase from chicken embryo fibroblast (CEF) cells. However, unlike the CEF kinase, it was not activated by treatment of cells with orthovandate or serum. Total S6 kinase activity in cell extracts was also stimulated by cAMP elevation, PMA, EGF, and insulin. None of these agents were additive with NGF in the activation of this S6 kinase. These results are unlike those seen in vivo or in the above in vitro studies. PMA did not stimulate either S6 phosphorylation in vivo or S6 kinase activity in vitro. Furthermore, among the same agents tested for additivity in stimulating S6 phosphorylation in vivo, only NGF and cAMP were not additive. The total NGF stimulated S6 kinase activity seen in cell extracts in the present study was not inhibited by the A-kinase inhibitor protein added to the extracts but was partially inhibited by prolonged phorbol ester pretreatment of cells. The S6 kinase activities described by BLENIS and ERIKSON (1986) are clearly not the same as the S6 kinase described by MATSUDA and GUROFF (1987). The former kinase activity is not inhibited by glycerophosphate and EGTA at levels which completely inhibit the S6 kinase activity described by MATSUDA and GUROFF. The characteristics for activation of the S6 kinase of MATSUDA and GUROFF best match those described for the rapid stimulation of S6 phosphorylation in PC12 cells (HALEGOUA and PATRICK 1980) and may well be the major kinase activity responsible for this NGF stimulated phosphorylation in vivo. It is possible that this S6 kinase activity predominates during an early phase (first 2 h) of S6 phosphorylation in response to NGF while the S6 kinase of BLENIS and ERIKSON is dominant during a later phase (up to 3 days) (HALEGOUA and PATRICK 1980). The involvement of S6 phosphorylation in the general increase in protein synthesis seen in response to NGF as well as the role of different modes of S6 phosphorylation in controlling S6 function have not yet been addressed.

Several other proteins phosphorylated in response to NGF have also been identified by in vivo labeling with $^{32}PO_4$. In one study (GREENE et al. 1983), increased labeling of several proteins was described in response to NGF. These included, in addition to TH, novel bands seen on SDS gels at apparent molecular masses of 300 kDa, 72 kDa, 64 kDa, and 56 kDa. Increased phosphorylation occurred over varying time periods, seen as early as 2 h and as late as 2 days and beyond. FGF also caused an increase in the labeling of the 300 kDa protein (RYDEL and GREENE 1987). The protein at 300 kDa was targeted for further study and was identified as a microtubule associated protein (MAP) designated MAP 1.2. Subsequent studies indicated that this protein undergoes extensive regulation by NGF in PC12 cells. It is rapidly phosphorylated within 15 min of NGF treatment (ALETTA et al. 1988a). Levels of the protein (GREENE et al. 1983; DRUBIN et al. 1985; ALETTA et al. 1988a), its mRNA (LEWIS et al. 1986), and the protein's phosphorylation state (ALETTA et al. 1988a) all increase over a period of several days. MAP 1.2 protein and mRNA levels increase about three- to fourfold, whereas the phosphorylation state increases up to about fourfold (ALETTA et al. 1988a). Other MAPs, termed chartins, are also phosphorylated in response to NGF (BLACK et al. 1986). The phosphorylation state of the chartins changes over a 2 week period of PC12 cell exposure to NGF such that the abundance of the acidic, highly phosphorylated variants increases relative to the basic, less phosphorylated variants. It is not clear whether these long-term changes in the phosphorylation state of MAPs are due to stimulated kinase activities, decreased phosphatase activities, or the relative accessibilities of these substrates to such enzymes. A role for the changes of both phosphorylation and relative abundance of these MAPs in microtubule assembly and neurite outgrowth has been repeatedly suggested and will be discussed further in Sect. 5.

A novel neuronal intermediate filament protein, peripherin, has recently also been found to be phosphorylated in response to NGF. Peripherin is present in cells in multiple forms, a subset of which is phosphorylated (ALETTA et al. 1989). NGF-stimulated phosphorylation of peripherin, evident in electrophoretic analysis of extracts from $^{32}PO_4$ labeled cells, is seen by 1 h and peaks after 2 h of NGF treatment. Peripherin is also phosphorylated in response to cAMP derivatives, the phorbol ester 12-*O*-tetradecanoylphorbol-13-acetate (TPA), or depolarizing levels of K^+, but not in response to EGF or insulin. The NGF-stimulated phosphorylation of peripherin still occurs in the A-kinase deficient PC12 cell variant (A126-1B2), indicating that A-kinase is not necessary for the major portion of this phosphorylation. The NGF-stimulated phosphorylation is seen also in PC12 cells which have been treated with high levels of TPA to down regulate C-kinase activity and in cells treated with dihydropyridine blockers of voltage dependent Ca^{++} channels. The authors suggested from these data that peripherin phosphorylation in response to NGF did not involve A-kinase, C-kinase, or CAMP kinase. However, problems with C-kinase down regulation approaches (discussed below) suggest that such experiments may not rule out the mediation of phosphorylations by this family of kinases. Furthermore, nitrendipines block only a specific class of voltage dependent Ca^{++} channels and do not rule out the involvement of Ca^{++} from other plasma membrane channels or released from intracellular stores. The involvement of C-kinase or CAMP kinase in peripherin phosphorylation thus remains an open question.

NGF-induced phosphorylation of synapsin I was examined using $^{32}PO_4$ labeled PC12 cells (ROMANO et al. 1987). Overall labeling of synapsin I was found to increase by about one-third within 30 min of treatment. Phosphopeptide mapping analysis indicated that the increased labeling occurred at one peptide designated peptide N. This peptide was distinct from those phosphorylated in vitro, using partially purified synapsin I, by any of four distinct protein kinases, A-kinase, CAMP kinases I and II, and C-kinase. On the basis of this result it was suggested that NGF uses a distinct kinase to carry out phosphorylation of synapsin I. It is not known whether a cascade mechanism involving, for example, C-kinase or CAMP kinase mediates this phosphorylating event or what relationships may exist between the in vitro phosphorylations and those carried out in vivo. However, cAMP elevation in PC12 cells did not elicit increased phosphorylation of synapsin I in spite of its phosphorylation by the A-kinase in vitro.

In a novel approach to the identification of functionally important kinases stimulated by NGF, LANDRETH and RIESER (1985) examined kinase activities which were associated with the PC12 cytoskeleton. Detergent-insoluble cell extracts were prepared in which the cytoskeleton (excluding microtubules) was well-preserved. The addition of [^{32}P]ATP to this preparation allowed the labeling of a large number of proteins. The labeling of one protein of 250 kDa was specifically increased in the extract from cells which had been pretreated with

NGF or EGF. The stimulation of kinase activity seen in situ occurred rapidly (maximal within 5 min) in response to the growth factors and transiently, dropping to near baseline after 30 min of continual treatment. The in vitro kinase activity was unaffected by cAMP, Ca^{++}, or trifluoperazine suggesting it was not A-kinase, C-kinase, or CAMP kinase (LANDRETH and RIESER 1985). Phosphorylation of pp250 was howcver, stimulated by phorbol ester treatment of the intact cells in a manner which was not additive with NGF (LANDRETH and WILLIAMS 1987). The ability to reproduce the phosphorylation of this protein in soluble extracts (LANDRETH and WILLIAMS 1987) should facilitate the purification and characterization of both kinase and substrate.

In three instances, decreased $^{32}PO_4$ labeling of proteins was reported in response to NGF. This was first reported for histone H2a (HALEGOUA and PATRICK 1980). It was not determined whether this was due to decreased kinase activity or increased phosphatase activity caused by the NGF treatment. Another protein, of 100 kDa, whose phosphorylation was similarly found to become decreased in response to NGF, was described in PC12 cells (END et al. 1982) and later in NGF responsive cultured neurons (HAMA and GUROFF 1985). This protein phosphorylation was also found to be decreased in response to FGF treatment of PC12 cells (TOGARI et al. 1985; RYDEL and GREENE 1987). Decreased labeling of this protein, designated NSP 100, was observed within 15 min of NGF treatment, and this effect was retained in cell free extracts from NGF treated PC12 cells. This decreased phosphorylation of NSP 100 in cell extracts (TOGARI and GUROFF 1985) was also seen in response to EGF, PMA, the adenosine receptor agonist NECA (END et al. 1983), and by elevation of intracellular Ca^{++} or cAMP (HASHIMOTO et al. 1986). Furthermore, in contrast to increased TH phosphorylation, decreased NSP 100 phosphorylation in response to NGF or cAMP elevation requires the presence of extracellular Ca^{++} (HASHIMOTO et al. 1986). NSP 100 and its kinase were partially purified and the NGF effect reconstituted (TOGARI and GUROFF 1985; HAMA et al. 1986). These studies concluded that the decreased phosphorylation of NSP 100 by NGF was mediated by an activation of C-kinase which then phosphorylated and inactivated a distinct NSP 100 kinase.

Several recent studies have attempted to evaluate the role of C-kinase in NGF-induced neurite outgrowth. Phorbol ester derivatives which stimulate C-kinase have been found to promote neurite outgrowth caused by NGF (CHANDLER and HERSCHMAN 1980). Two recent studies have attempted to specifically eliminate C-kinase activity from PC12 cells to test the potential requirement of C-kinase in NGF-induced neurite outgrowth. In one case, sphingosine, a drug reported to inhibit C-kinase in vitro and in vivo in some cell types was used (HALL et al. 1988). Treatment of cells with relatively high doses of sphingosine was reported to block NGF-induced neurite outgrowth in a reversible manner. The drug was also found to inhibit NGF stimulated incorporation of $^{32}PO_4$ into a wide array of proteins seen on one-dimensional gels, a result which was suggested to indicate the inhibition of C-kinase. However, the specific enhanced incorporation of label into TH by NGF was still seen in the presence of the drug. The authors concluded that the activation of C-kinase by NGF was required for neurite outgrowth. In our laboratory, sphingosine, over a wide range of concentrations, was not able to inhibit either phorbol ester-induced or NGF-stimulated phosphorylation of TH in PC12 cells (R. ARMSTRONG and S. HALEGOUA, unpublished work). At high concentrations of sphingosine, such as those used in the above study, the cells became detached from the culture dish, calling into question inhibitory effects of this drug. In a subsequent study, prolonged treatment with phorbol esters was used to down regulate C-kinase in PC12 cells (YOUNG et al. 1987). Long-term treatment of PC12 cells with phorbol dibutyrate (PDBU) was found to decrease C-kinase activity levels by about 90% (REINHOLD and NEET 1989), with no effect on NGF-induced neurite outgrowth. The authors concluded that C-kinase was not essential to this NGF action. However, in view of both the variety of C-kinase isoforms that exist, which have different sensitivities to phospholipids, diacylglycerols and phorbol esters, and their different subcellular locations, negative results from down regulation experiments must be viewed with caution. For example, in our laboratory, down regulation with PMA was able to block TH phosphorylation normally stimulated by this phorbol ester. However TH phosphorylation was stimulated by PDBU or dioctanoylglycerol (DOG) in these same PMA down regulated cells (J. CREMINS, R. ARMSTRONG, and S. HALEGOUA unpublished work). Identification of the C-kinase isoforms and their relative abundance and subcellular locations in PC12 cells must be studied further before definitive conclusions can be drawn from the above studies.

A summary of second messenger pathways, their interactions, and protein phosphorylation events which have in some ways been linked to these pathways is presented in Fig. 3. It is abundantly clear that NGF action involves changes in the phosphorylation state of a variety of different proteins. These changes include increased and decreased phosphorylations mediated through enhanced or reduced kinase activities and appear to include changes in phosphoprotein phosphatase activities as well. The phosphorylation events occur during a broad time range after NGF addition and in many cases involve kinase cascades. A number of the phosphorylated proteins have been identified and in one case thus far the phosphorylation of TH resulted in activation of the enzyme (and increased catecholamine synthesis). Other identified kinase substrates have functions more difficult to assess, but the correlation of phosphorylation with NGF treatment and actions suggests a physiological importance. Although most of the kinase actions identified to date involve modifications of serine and threonine residues, phosphorylations on tyrosines have also recently been described (MAHER 1988). As discussed below, the tyrosine kinase activities are rapidly activated and are likely to be mediators of signal transduction for NGF. As illustrated in Fig. 3, some second messenger pathways stimulated by NGF may not involve protein phosphorylations. Their mode of action remains to be determined.

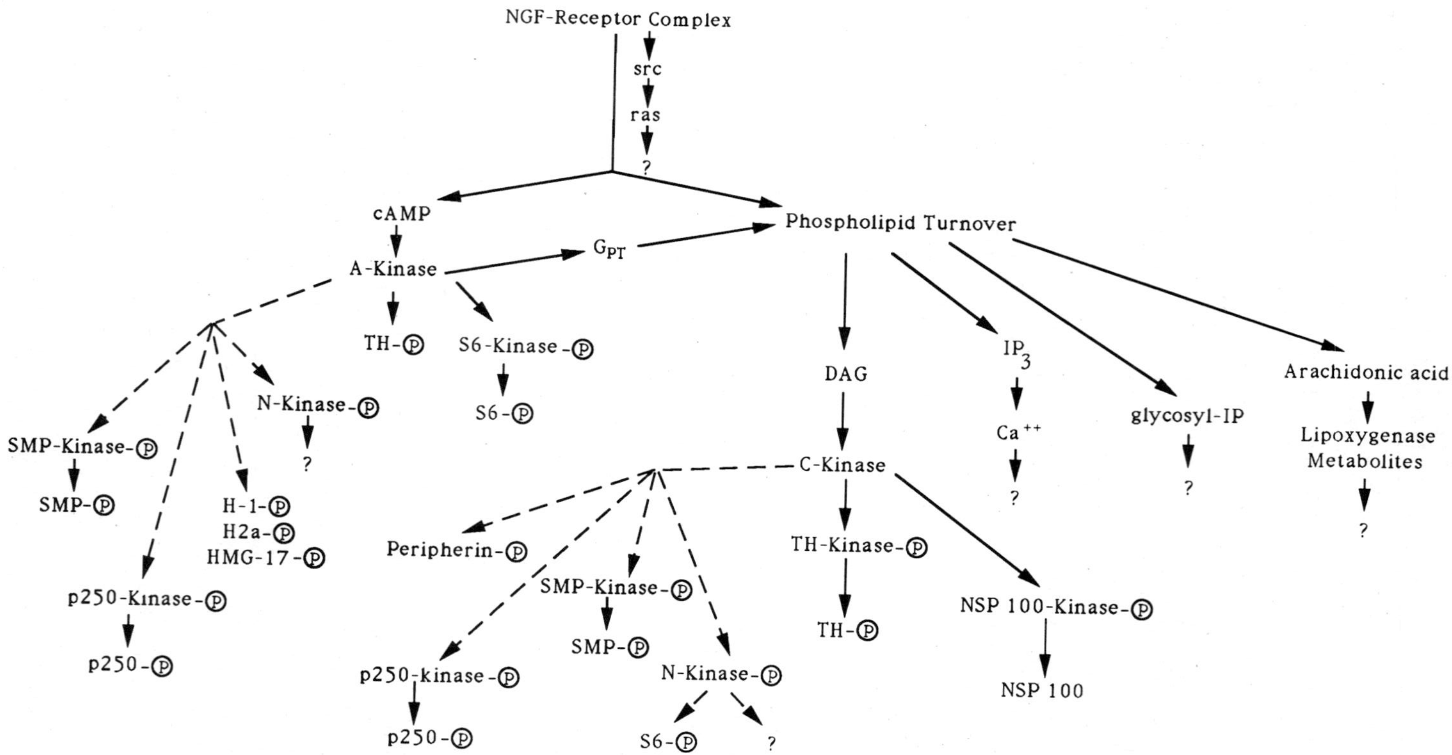

Fig. 3. Second messenger pathways stimulated by NGF and some of the resulting protein phosphorylation events in PC12 cells. A variety of second messenger pathways stimulated by the NGF receptor complex is shown. At least a subset of these are presumed to be carried out through the *src* and *ras* signal transduction molecules. Only those protein phosphorylations which have been associated with second messenger pathways, either by correlative studies (shown by *dashed arrows*) or by more direct experimental evidence (shown by *solid arrows*) are shown. Studies are reviewed in text

3.5 Proto-Oncogenes

From the studies above it is apparent that multiple second messenger pathways may be involved in the various actions of NGF. The generation of these second messenger molecules result from the binding of NGF to its receptor, the sole product of a single gene. It appears that a single, high affinity receptor is responsible for mediating all NGF actions (see CHAO, this volume). As the NGF receptor, unlike other growth factor receptors, has no known intrinsic enzymatic activity, it must apparently interact with at least one other signal transducing molecule. The identification of such companion molecules has only recently been undertaken.

The guanyl nucleotide binding proteins (G-proteins) (for review see GILMAN 1984; NEER and CLAPHAM 1988) are the only class of signal transducers known to be linked to a variety of receptors including those for hormones, neurotransmitters, and light. The G-proteins become activated and bind guanosine triphosphate (GTP) after linking to ligand-occupied receptors and inactivated after hydrolyzing bound GTP (intrinsic GTPase activity). The best characterized G-protein is G_s, which when activated in response to ligand binding, stimulates adenylate cyclase. Since NGF action appears to involve the production of cAMP and the activation of A-kinase, the ability of NGF to stimulate adenylate cyclase has been examined (RACE and WAGNER 1985). In this study, adenylate cyclase activity in vivo was determined by measuring the rate of [^{3}H]cAMP production from [^{3}H]adenosine. Although NGF treatment did not stimulate [^{3}H]cAMP production, isoproterenol (which acts through the G_s linked β-adrenergic receptor) stimulated [^{3}H]cAMP production by tenfold. This study raised the possibilities that cAMP accumulation by NGF might be due to either phosphodiesterase inhibition or that NGF might use an alternative means of activating A-kinase. Thus, the potential involvement of G_s or other pathways for A-kinase activation by NGF remain to be elucidated. Although no simple means of directly inhibiting G_s exists, pertussis toxin (PT) catalyzes the ADP-ribosylation and resulting inactivation of several G-proteins, including those which mediate inhibition of adenylate cyclase and stimulate PI turnover. As discussed above, PT blocks that portion of NGF-induced PI turnover which is carried out through a cAMP dependent pathway. However, the majority of the NGF stimulated PI turnover is not blocked by PT and is thus not carried out by PT sensitive G-proteins (J. CREMINS and S. HALEGOUA, unpublished work). Furthermore, prolonged treatment of PC12 with PT before and during NGF treatment was only slightly inhibitory for neurite outgrowth (S. HALEGOUA, unpublished work). These negative results stimulated the search for other potential G-protein-like molecules as NGF signal transducers.

The *ras* gene family encodes membrane bound proteins with GTP binding and hydrolyzing activity (BARBACID 1987). It has been suggested that these proteins play a role analogous to G-proteins in mediating cell proliferation by growth factors and hormones, because "activated" (i.e., viral or mutant) forms of these proteins promote cell proliferation (BARBACID 1987) and because their inactivation, such as by microinjection of antibodies to ras proteins, blocks these growth promoting effects (Smith et al. 1986). Surprisingly, however, the introduction of Ha-*ras* or N-*ras* oncogenes into PC12 cells by viral infection (Noda et al. 1985), transfection (Guerrero et al. 1986), or microinjection of the oncogenic Ha-ras protein (Bar-Sagi and Feramisco 1985) resulted in the morphological differentiation of PC12 cells in the absence of NGF. This result suggested that the oncogenic, activated ras proteins may couple to the same intracellular signaling pathways as the occupied NGF receptor. For example, when the N-*ras* oncogene in a retroviral vector was transfected into PC12 cells it resulted in the extensive outgrowth of neurities, similar to those produced by NGF (Guerrero et al. 1986). However, the morphological effects of N-*ras* differed from that of NGF in that N-*ras* caused an unusual somatic enlargement followed by cell degeneration within about 2 weeks after transfection. In a subsequent study, the v-Ha-*ras* gene, engineered under the control of the mouse mammary tumor virus (MMTV) promotor, was stably transfected into PC12 cells and induced with dexamethasone (Sugimoto et al. 1988). In this system, both v-Ha-*ras* and NGF similarly caused Li^+-inhibitable neurite outgrowth, transcription dependent priming of the cells for subsequent rapid neurite outgrowth by NGF (see below), survival in serum free medium, and increases of acetylcholinesterase activity. Induction of N-*ras* in PC12 transfectants resulted in the induction of ODC (Guerrero et al. 1988), as did treatment of PC12 cells with NGF. However, the characteristic rapid induction of the c-*fos* gene in response to NGF was not observed after N-*ras* induction, leading to the suggestion that ras p21 does not mediate this NGF effect. This result is surprising in light of the fact that oncogenic mutants of Ha-*ras* or N-*ras* are able to cause induction of the endogenous c-*fos* gene in fibroblasts (Stacey et al. 1987) and of c-*fos* promoter-chloramphenicol acetyltransferase (CAT) constructs transfected into PC12 cells (Sassone-Corsi et al. 1989). However, the initial induction of MMTV-N-*ras* by dexamethasone is slow (occurring over many hours) relative to the rapid (within 5 min) and transient (complete within 2 h) induction of c-*fos* seen in response to acute activators such as NGF. It is thus possible that N-*ras*-mediated induction of c-*fos* was obscured by the rapid, transient, and refractory nature of c-*fos* gene regulation (see Sect. 4.1).

An independent study has addressed the possible role of the cellular ras p21 proteins in mediating one specific aspect of NGF action, neurite outgrowth. In this study, a monoclonal anti-ras p21 antibody was microinjected into PC12 cells in order to block ras p21 protein function (Hagag et al. 1986). To eliminate the deleterious effects of micropipette impalement on the small PC12 cell, these cells were first fused to create NGF responsive, giant multinucleated cells (O'Lague and Huttner 1980). Microinjection of anti-ras p21 antibody into the fused PC12 cells completely blocked neurite outgrowth by NGF (Hagag et al. 1986) and FGF (N. Kremer, J. Brugge and S. Halegoua, unpublished work). These antibody results have recently been confirmed by the introduction into PC12 cells of an inactive but competing mutant Ha-*ras* (Feig and Cooper 1988). Taken together, the above results suggest that the product of a member of the *ras* gene family either directly or indirectly mediates the coupling of NGF receptors and FGF receptors to the intracellular mediators of neurite outgrowth. Introduction of v-Ha-*ras* p21 into cultured neurons has recently been found to alleviate their dependence on NGF, brain-derived neurotrophic factor (BDNF), and ciliary neurotrophic factor (CNTF) (DOMENICO BORASIO et al. 1989) further suggesting a common role for *ras* in mediating neuronal growth factor action.

One of the limitations in understanding the role of *ras* in NGF action is that the understanding of *ras* in other physiological processes and in oncogenesis is still incomplete. In yeast cells the *ras* genes are coupled to adenylate cyclase (TODA et al. 1985). However, this function has not been seen in mammalian cells. Several reports have connected ras protein activity with the generation of PI metabolites. The *ras* oncogene products were found to result in the increased production of PI metabolites, particularly diacylglycerol (see for example FLEISCHMAN et al. 1986). However, as with NGF, much of the diacylglycerol production by ras proteins has been suggested to be derived from lipids other than PI (LACAL et al. 1987). Microinjection of Ha-ras p21 proteins has also been shown to rapidly stimulate phospholipase A2 activity (BAR-SAGI and FERAMISCO 1986). Since NGF was found to stimulate protein phosphorylation caused by stimulation of A-kinase as well as C-kinase (see above), it was suggested that ras p21 proteins may mediate stimulation of either or both of these pathways (HAGAG et al. 1986). In support of this hypothesis, SUGIMOTO et al. (1988) have found that v-Ha-*ras* induction in PC12 resulted in the stimulation of cAMP production and PI turnover. Furthermore, *Ha-ras* oncogenes were found to cause gene inductions through the serum response element of the c-*fos* gene and through the phorbol ester response element (TRE) in PC12 cells (SASSONE-CORSI et al. 1989). However, the long lag time for production of PI and cAMP metabolities in these cells (SUGIMOTO et al. 1988) are difficult to reconcile with ras protein acting as an immediate coupler to these pathways. An additional complexity has been presented by YU et al. (1988), who suggested that ras proteins act at a step subsequent to PI turnover in mediating proliferation responses in 3T3 cells. The coupling of NGF receptors to intracellular mediators, such as cAMP and PI turnover, through a ras protein thus remains an intriguing hypothesis. However, it is likely that ras is only one component of a novel and complex signal transduction pathway for NGF action (see below).

One class of receptors with an intrinsic signal transduction capability is that of the mitogenic growth factors. These receptors contain a carboxy terminal cytoplasmic tyrosine kinase domain which is activated by ligand binding and which is necessary for signal transduction. Although NGF receptor does not appear to contain tyrosine kinase activity, several lines of evidence point to a role for tyrosine kinase (and possibly a src kinase) in mediating NGF action. The first indication that tyrosine kinase activity may play a role in neuronal differentiation came from infection of PC12 cells with Rous Sarcoma Virus (ALEMÁ et al. 1985). Although infection of most cell types by this virus results in carcinogenic transformation, the viral infection of PC12 cells resulted in morphological differentiation via the src kinase, similar to that produced by NGF treatment. To further examine the differentiation of PC12 cells induced by v-*src*, cloned v-*src* mutants which encode a temperature sensitive v-*src* (*ts* v-*src*) kinase activity was stably transfected into PC12 cells (RAUSCH et al. 1989; S. THOMAS, M. HAYES, G. D'ARCANGELO, R. ARMSTRONG, M. DEMARCO, J. BRUGGE, and S. HALEGOUA, unpublished work). In such PC12 transfectants, shifts in the growth temperature control both *src* activity and *src*-induced differentiation (*ts* v-*src* is active at 35 °C and inactive at 41 °C). The use of these cell lines has demonstrated that *src*-induced and NGF-induced differentiation share many characteristics besides neurite outgrowth. The differentiation is reversible, proceeds long-term (over weeks), is able to prime subsequent rapid neurite regrowth by NGF, and includes the induction of genes also stimulated by NGF such as c-*fos*, *NGFIa, thy 1, SCG10*, and *ODC*. In contrast to NGF, v-*src* induced neurite outgrowth is cell density independent (S. THOMAS et al., unpublished work) and tetrodotoxin (TTX)-sensitive Na^+ channels are not induced, suggesting that these effects may reflect unique steps in the NGF signal transduction cascade.

The above results raise the interesting question of whether c-src, or another tyrosine kinase, is involved in differentiation by NGF. This issue has been addressed in two ways. MAHER (1988) assessed tyrosine kinase activity in PC12 cells by performing Western blotting of PC12 proteins and probing with anti-phosphotyrosine antibodies. She found that NGF treatment of cells rapidly (within 2 min of treatment) resulted in increased binding of these antibodies to specific protein bands in the blot, suggesting that NGF stimulated protein phosphorylations on tyrosine residues. The pattern of proteins was similar but not identical to that obtained in response to EGF treatment of the cells. Unlike EGF-stimulated tyrosine phosphorylations by the EGF receptor kinase, those stimulated by NGF were attenuated by low temperature. These results and the fact that the NGF receptor cDNA does not appear to encode a tyrosine kinase were taken to imply that NGF receptor activates a tyrosine kinase with which it may interact. The same anti-phosphotyrosine antibodies used in immunofluorescence staining of PC12 cells illuminated components which were correlated with the localization of actin filaments particularly at the neurite growth cone (MAHER 1989), where c-src (SOBUE and KANDA 1988) and NGF receptors are also localized in PC12 cells (LEVI et al. 1980).

In a set of experiments carried out recently in our laboratory, the potential requirement for c-src kinase in NGF action has been assessed. Microinjection of a specific monoclonal anti-src antibody (MAb 327) has been found to inhibit v-src-stimulated DNA synthesis in fibroblasts (J. WALTON, J. FERAMISCO and J. BRUGGE, unpublished work). We have found that microinjection of MAb 327 into fused PC12 cells stably expressing the *ts* v-*src* gene prevented neurite outgrowth at the permissive temperature (N. KREMER, J. BRUGGE, and S. HALEGOUA, unpublished work). Microinjection of these antibodies into fused PC12 cells also resulted in the block of NGF-induced neurite outgrowth and could cause retraction of already extended neurites. These results indicate that src (or an unidentified closely related antigen) plays an essential role in NGF-induced neurite outgrowth. Whether c-src is the kinase responsible for the observed NGF stimulated tyrosine phosphorylations remains to be determined.

From the above studies it is apparent that members of both the c-*src* and c-*ras* gene families may play essential roles in mediating NGF action. These results are most easily interpreted by a model in which NGF results in the stimulation of both src and ras p21 to act sequentially in a signal transduction pathway. Microinjection studies in fibroblasts have shown that anti-ras antibodies block v-src-induced transformation. This result suggests a preferred order for src and ras functions in NGF-induced differentiation with src occurring first, and ras activity occurring downstream. Further support for this model comes from additional microinjection experiments (N. KREMER, J. BRUGGE, and S. HALEGOUA, in preparation). Injection of anti-ras antibody into *src*-transfected PC12 cells blocks v-src-induced neurite outgrowth. Importantly, the converse is not true. Injection of anti-src antibody does not inhibit neurite outgrowth from dexamethasone-treated PC12 cells stably expressing an MMTV-*ras* fusion gene.

The results discussed above underscore the striking similarities between src and ras functions in mitogenesis and neuronal differentiation. Although the molecular details of the oncogene relationships are at present unclear, a clue may lie in the GTPase activating protein (GAP) which can associate with ras p21. GAP has recently been identified as a major substrate for tyrosine kinases, (ELLIS et al. 1990). Thus, tyrosine phosphorylation by src and the resultant inactivation of GAP could be one means of elevating ras p21 activity.

While these results support the above model in which a src-like tyrosine kinase mediates NGF-induced neurite outgrowth through a ras-related G-protein, it does not preclude differentiating effects of src which may occur independently of ras p21 or of NGF receptor acting independently of src or ras p21. A comparative analysis of various markers of differentiation induced by NGF and by the *src* and *ras* oncogenes combined with antibody blocking experiments may further reveal multiple pathways of NGF-induced differentiation. These experiments together with those assessing the various second messenger activities stimulated by NGF (see above and Fig. 3) may shed light on the specific transduction pathways which underlie NGF actions.

3.6 Pharmacological Inhibitors

Interference with NGF signal transduction by drugs has provided an approach toward identification of target transducing molecules. Drugs which can inhibit selectively NGF action, but not other receptor-mediated events, have been particularly useful. The first class of drugs with this property are the inhibitors of methylation reactions, such as methylthioadenosine (MTA). Such drugs were found to block NGF-induced neurite outgrowth but not EGF actions (SEELEY et al. 1984), suggesting a role for either protein or phospholipid methylation in NGF action. This finding is consistent with earlier data showing that NGF stimulated phospholipid methylation in primary neuronal cultures (PFENNINGER and JOHNSON 1981; SKAPER and VARON 1984). The importance of phospholipid methylation was examined more carefully in PC12 cells by FERRARI and GREENE

(1985). Using [^{3}H]methyl-methionine labeling of PC12 cultures, these investigators saw no effect of NGF on phospholipid labeling. Furthermore, concentrations of methylation inhibitors which blocked phospholipid methylation had no effect on NGF-induced neurite growth. The results therefore suggested that the inhibition of neurite growth by methylation inhibitors occurred at concentrations more consistent with the inhibitors' ability to inhibit protein methylation. Given the higher concentrations of MTA required for neurite inhibition, it is also possible that other reactions might be the critical ones inhibited. For example, recent reports have indicated that MTA is a potent inhibitor in vitro and in vivo of both an NGF-stimulated pp 250 protein kinase (SMITH et al. 1989) and the tyrosine kinase stimulation produced by NGF treatment (MAHER 1988). Because similar events induced by EGF are not blocked, it suggests that the drug target may lie close to the NGF receptor in the signal transduction pathway. The differentiating action of FGF on PC12 cells is also blocked by MTA, indicating that a receptor-mediated event common to NGF and FGF is affected by the drug. Whether the essential site of action of MTA is a kinase, methyltransferase, or other cellular component remains to be determined.

Another drug of interest which has more recently been described is K252a. This drug was first discovered to inhibit C-kinase both in vivo and in vitro. It has since been found to inhibit a variety of protein kinases. When tried on PC12 cells, K252a was found to be a highly specific inhibitor of NGF action. It appears to block every NGF action thus far tested, from tyrosine kinase and serine kinase stimulations to gene inductions and neurite outgrowth (HASHIMOTO 1988; KOIZUMI et al. 1988; CHO et al. 1989; LAZAROVICI et al. 1989; MAHER 1989; SMITH et al. 1989). No similar actions by FGF or EGF appear to be affected by the drug. It has thus been suggested that K252a blocks an event specific to NGF receptor action, and it has also been suggested that it may act by inhibiting the tyrosine kinase stimulated by NGF (MAHER 1989). However, v-*src* induced neurite outgrowth is unaffected by the drug (RAUSCH et al. 1989). Since it does not affect NGF binding to receptor, K252a may block an initial transduction event, perhaps a kinase activation. However, as with MTA, the essential site(s) of action for K252a have yet to be identified.

4 Regulation of Gene Expression

Our present model for NGF-induced differentiation (Fig. 1) invokes major changes in gene expression. The first suggestion that changes in gene expression were important in NGF action came from studies on NGF-induced neurite outgrowth. After NGF addition to PC12 cells, there is a characteristic lag time of about 48 h before neurite outgrowth is observed. Inhibition of RNA synthesis during this lag time results in a block of neurite outgrowth. The use of camptothecin at a level which blocks only 30% of total RNA synthesis completely blocks neurite outgrowth by NGF. Cells which are treated with NGF for periods longer than the lag time can rapidly regenerate neurities after replating with NGF, even in the absence of RNA synthesis. Elicitation of the capability for precocious neurite growth in the absence of RNA synthesis has been termed "priming" (GREENE et al. 1982). FGF can also prime PC12 cells: FGF-primed PC12 cells will elicit neurite regeneration in response to subsequent treatment with either NGF or FGF (TOGARI et al. 1985; RYDEL and GREENE 1987). These data led to the conclusion that neurite outgrowth by NGF and FGF was mediated by the transcriptional activation of gene(s) encoding proteins required for neurite outgrowth.

The first studies of NGF-induced alterations in protein synthesis changes were of a general nature and used one- and two-dimensional gel electrophoresis to resolve the protein species (MCGUIRE et al. 1978; GARRELS and SCHUBERT 1979; MCGUIRE and GREENE 1980; TIERCY and SHOOTER 1986). Using these techniques, only small quantitative changes in the synthesis of proteins in response to NGF were found. Using two-dimensional gel electrophoresis, incorporation of [^{3}H]fucose and [^{3}H]glucosamine into a 230 kDa protein termed NGF inducible large external glycoprotein (NILE) was found to be stimulated by NGF after 2 days of treatment. NILE synthesis, but not its glycosylation, was stimulated about threefold by NGF (MCGUIRE et al. 1978) and FGF (RYDEL and GREENE 1987). The increased synthesis by NGF was blocked by camptothecin, suggesting a requirement for RNA synthesis. The potential role of NILE in PC12 adhesion and in neurite outgrowth will be discussed in Sect. 5. GARRELS and SCHUBERT (1979) quantitatively analyzed the synthesis of over 800 proteins by two-dimensional gel electrophoresis of proteins labeled with [^{35}S]methionine in PC12 cells after 2 days of NGF treatment. Among these proteins, NGF treatment resulted in from 30% to tenfold changes in a number of proteins but no qualitative changes in protein synthetic rates. A more recent study of PC12 protein synthesis using two-dimensional gel electrophoresis focused on changes during the first few hours of NGF treatment (TIERCY and SHOOTER 1986). Several quantitative increases in synthetic rates of proteins were found which were blocked by the transcription inhibitor, actinomycin D. One study (MCGUIRE and GREENE 1980) using one-dimensional SDS-PAGE to analyze [^{35}S]methionine labeled PC12 proteins concentrated on an as yet unidentified protein of 80 kDa (p80) whose synthesis was found to be maximally stimulated by NGF after 2–3 days of treatment. The induction and maintenance of p80 synthesis by NGF was found to be selectively sensitive to camptothecin.

More directed approaches have been aimed at identifying proteins whose syntheses are regulated by NGF to promote the outgrowth of neurities. Included in this class are MAPs, including the chartin proteins (BLACK et al. 1986), tau proteins, and tubulins (DRUBIN et al. 1985), the synthesis of all of which are increased by NGF. The regulation of PC12 intermediate filament proteins by NGF has also been approached using diverse methods. In immunocytochemical studies using monospecific antibodies against each of the three neurofilament

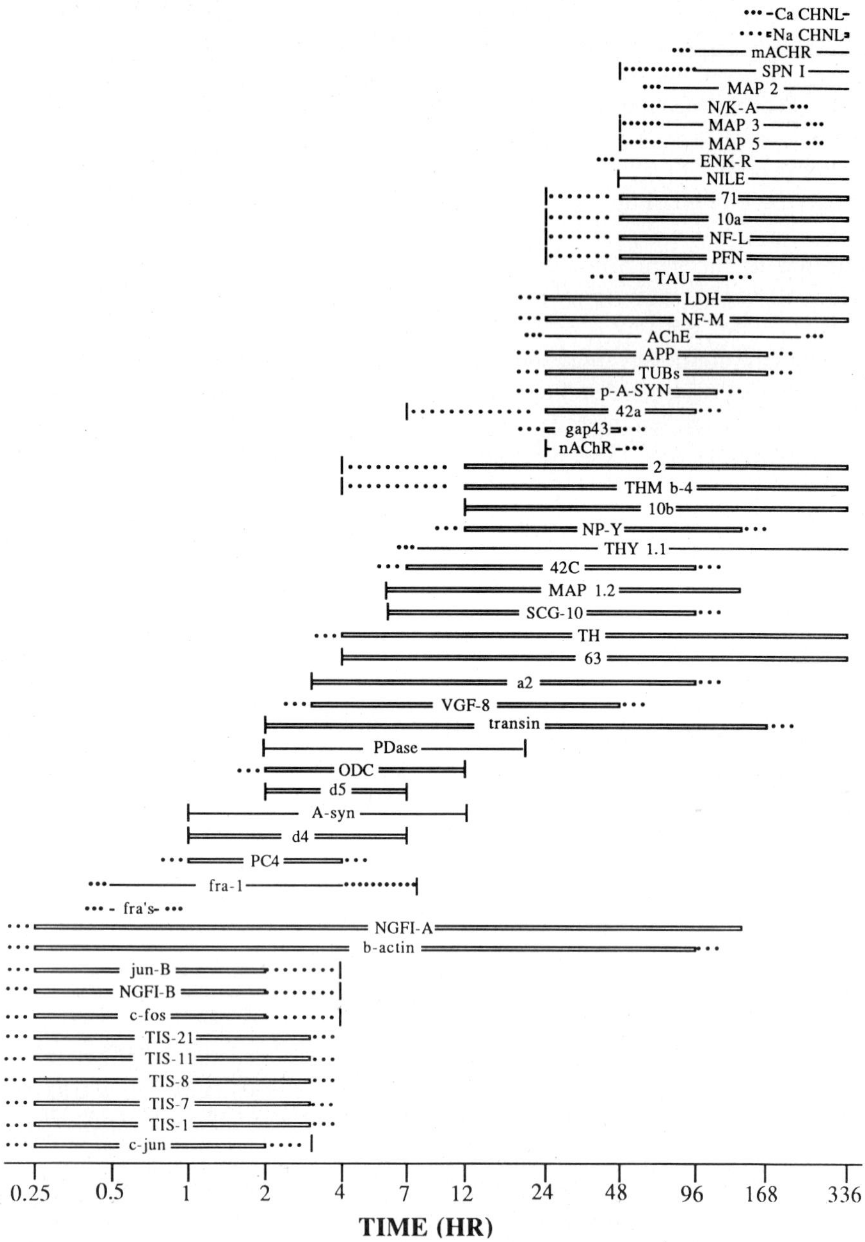

Fig. 4. Time course plot of NGF-induced changes in gene experssion in PC12 cells. Only those changes in gene expression which have been determined for identified proteins or mRNAs and for more than one time point after NGF treatment of cells are plotted. Changes in protein levels or protein synthesis (*solid lines*) and in mRNA levels (*open bars*) are shown. In cases where both types of data exist, only the data for mRNA levels are shown. Undetermined extensions of time courses are shown by *dots*; determined time points where no change from control (no NGF treatment) was found is shown by *vertical lines*. Times after NGF treatment are plotted on a logarithmic scale.

(*Continued*)

subunit proteins and vimentin, increased cell staining suggested that the abundance of these proteins was increased after NGF treatment (LEE et al. 1982; LEE and PAGE 1984). Some of these findings were confirmed and extended using one- and two-dimensional gel electrophoresis and immunoprecipitation (LEE 1985; LINDENBAUM et al. 1987). The increased production of the three neurofilament protein subunits (NF-L, NF-M, and NF-H) were found to be differentially regulated by NGF, with NF-H being increased much more slowly and to a lesser extent. Since vimentin is not detected in PC12 cells by two-dimensional gel analysis, cell staining with the anti-vimentin antibodies was likely due to cross-reactivity with peripherin (PORTIER et al. 1982, 1983). This protein was found to be induced by NGF in PC12 cells (PARYSEK and GOLDMAN 1987). The molecular cloning of cDNAs for tubulin, MAPs and peripherin and the use of these cDNAs to study regulation of these genes by NGF will be discussed below in Sect. 4.2.

Other directed approaches have included assays for specific enzyme activities such as choline acetyltransferase (SCHUBERT et al. 1977; GREENE and REIN 1977), TH (GOODMAN and HERSCHMAN 1978), acetylcholinesterase (AChE) (see for example LUCAS et al. 1980; RIEGER et al. 1980), ODC (see for example HATANAKA et al. 1978; HUFF and GUROFF 1979), Na^+/K^+ ATPase (INOUE et al. 1988), neuron specific enolase (VINORES et al. 1981), 2′–5′ oligo A synthetase, and 2′ phosphodiesterase (SAARMA et al. 1986) which are all stimulated by NGF treatment. ODC (TOGARI et al. 1985; RYDEL and GREENE 1987) and AChE (RYDEL and GREENE 1987) levels are also increased in response to FGF. These increases in activity have generally been shown to be due at least in part to increases in protein levels of the enzymes. Curiously, the enzyme lactate dehydrogenase was described as a single stranded DNA binding protein whose protein and mRNA levels are reduced by NGF; the relevance of this observation is thus far speculative (BIOCCA et al. 1984; CALISSANO et al. 1985). Levels of poly ADP-ribose synthetase are also reduced by NGF treatment (TANIGUCHI et al. 1988). In addition to enzymes, several neuropeptide transmitters including neurotensin (interestingly, glucocorticoid is permissive for NGF induction of neurotensin [TISCHLER et al. 1982]) and neuropeptide Y (ALLEN et al. 1984) and neurotransmitter receptors such as acetylcholine receptors (JUMBLATT and TISCHLER 1982; MITSUKA and HATANAKA 1984) and enkephalin receptors (INOUE and HATANAKA 1982) have been found to be induced by NGF in PC12 cells. By contrast, the EGF receptor is repressed, which may play a role in the NGF inhibition of cell division (LAZAROVICI et al. 1987). Secreted forms of the β-amyloid precursor (SCHUBERT et al. 1989) and levels of synapsin I (ROMANO et al. 1987) are also increased by NGF treatment.

To Fig. 4.
Abbreviations not defined in text are: *b-actin*, beta actin; *A-syn*, 2′-5′-A synthetase; *PDase*, 2′-phosphodiesterase; *NP-Y*, neuropeptide Y; *THM b-4*, thymosin beta 4; *nAChR*, nicotinic acetylcholine receptor; *p-A-syn*, poly (ADP-ribose) synthetase; *TUBs*, alpha and beta tubulins; *AChE*, acetylcholine-esterase; *LDH*, lactate dehydrogenase; *PFN*, peripherin; *ENK-R*, enkephalin receptor; *N/K-A*, sodium/potassium ATPase; *SPN I*, synapsin I; *mAChR*, muscarinic acetylcholine receptor; *Na CHNL*, sodium channel; *Ca CHNL*, calcium channel

The changes in electrical properties in PC12 cells treated with NGF have included the induction of both voltage dependent Na^+ channels (DICHTER et al. 1977) and Ca^{++} channels (see for example TAKAHASHI et al. 1985; GARBER et al. 1989). Complementary DNA clones are now available for a number of the above proteins and studies of mRNA regulation by NGF using these clones will be discussed below.

Molecular studies of NGF action have focused on identifying components of two classes of genes. One class encodes proteins which contribute directly to the differentiated sympathetic neuronal phenotype, such as those for proteins which are components of neuritic processes. The other class of genes contributes indirectly to the neuronal phenotype by causing the activation of downstream genes which are directly involved in NGF actions. What has emerged from these studies is a precise time sequence of gene inductions occurring over an interval spanning 5 min to several days of continuous NGF treatment (Fig. 4). Studies of the induced genes, their mode of induction, and their relationships to each other and to the establishment of neuronal characteristics are making valuable contributions to a molecular framework of NGF action.

4.1 Primary Response Genes

Parallels have been drawn between cellular genes induced by growth factors and genes induced upon viral infection. In virology, "immediate early genes" were first defined as those genes encoded by viruses which are expressed very early and only transiently during viral infection of mammalian cells. The induction of these genes upon viral infection is unique in that they do not require ongoing protein synthesis but rather are "superinduced" in the presence of protein synthesis inhibitors, due in part to stabilization of their mRNAs. In the extrapolation of the viral cycle to a model for growth factor action (see Fig. 1) a set of cellular genes are rapidly (within minutes) induced by growth factors and can be superinduced when protein synthesis is blocked. Genes of this class (primary response genes), induced by growth factors and other mitogenic stimuli, have been identified and cloned. Primary response gene induction is not restricted to mitogenic stimuli but occurs in differentiating and other modulatory systems. NGF-induced neuronal differentiation of PC12 cells also falls into this category. The rapidity with which primary response genes are turned on suggests that their induction is likely to be a direct result of second messenger action. In general, a role of primary response genes as transcriptional trans-activators has been suggested to provide a link between rapid second messenger stimulation and long-term gene induction.

The first primary response gene to be characterized was the proto-oncogene c-*fos*, the cellular counterpart of the transforming v-*fos* gene of two murine retroviruses (for review see CURRAN 1988). The c-*fos* gene is turned on by polypeptide growth factors which stimulate mitogenesis, in cells which are induced to differentiate, and in post-mitotic neurons which are electrically stimulated (MÜLLER et al. 1984). PC12 cells have provided an ideal system for studying the molecular basis of all of these paradigms of c-*fos* induction.

Stimulation of at least three major second messenger pathways can lead to c-*fos* induction in PC12 cells. Activation of C-kinase by phorbol esters or diacylglycerol derivatives, cAMP elevation by chemical derivatives of cAMP or by cholera toxin, and elevation of intracellular free Ca^{++} by membrane depolarization or by the Ca^{++} ionophore A23187 have all been shown to induce c-*fos* (KRUIJER et al. 1985; GREENBERG et al. 1985; SHENG et al. 1988; R. ARMSTRONG and S. HALEGOUA, unpublished work). In addition, stimulation of the tyrosine kinase EGF receptor or the putative tyrosine kinase FGF receptor (GREENBERG and ZIFF 1984), and the v-src tyrosine kinase (FUJII et al. 1989; R. ARMSTRONG and S. HALEGOUA, unpublished work) induce c-*fos*. In all of these cases, as with NGF, the kinetics of c-*fos* induction are rapid and transient, with gene transcription beginning within 5 min of the stimulus, peaking at 15–30 min, and returning to basal levels within 1 h.

Multiple genetic elements found in the upstream region of the c-*fos* gene are responsible for its induction. One element with dyad symmetry has been identified which causes the induction of a reporter gene such as bacterial CAT to which it has been linked (TREISMAN 1985). The dyad-reporter fusion gene is induced in PC12 cells in response to C-kinase activation, NGF (SHENG et al. 1988), and other growth factors. The fusion gene is also induced in response to serum stimulation of fibroblasts. The dyad symmetry element is contained in what has been termed a serum response element (SRE). Recently, a second NGF responsive element has been identified which lies immediately downstream of the dyad symmetry element (VISVADER et al. 1988). This element is capable of conferring NGF responsiveness to a heterologous promoter in the absence of the dyad containing SRE, and mediates reporter gene induction in response to serum stimulation in fibroblasts. Although each of these elements confers NGF responsiveness independently, the presence of both elements does not result in an additive c-*fos* induction. This second NGF and SRE has been termed SRE-II, as opposed to the dyad symmetry-containing element now termed SRE-I.

Another distinct DNA element capable of mediating c-*fos* induction is the cAMP responsive element (CRE) consensus sequence present in a variety of eukaryotic genes (COMB et al. 1986; MONTMINY et al. 1986). Like other CREs, the c-*fos* CRE is capable of mediating cAMP induction when linked to a reporter gene (SASSONE-CORSI et al. 1988), however, other regions in the c-*fos* gene may contribute to the cAMP induction of c-*fos* (FISCH et al. 1989). There are presently no data suggesting a role for CREs in NGF-mediated induction of c-*fos*. Greenberg and colleagues (SHENG et al. 1988) have reported that the CRE is responsible for c-*fos* induction by Ca^{++}. Whether the induction is due to a Ca^{++}-mediated increase in cAMP or to a convergence on a common activating protein has not yet been addressed. However, evidence discussed below suggests that cAMP-mediated and Ca^{++}-mediated induction of c-*fos* do not share a common pathway. Two other regulatory elements identified in the upstream region of c-*fos* include one which is responsive to v-sis conditioned media treatment of fibroblasts and another which is an AP1 binding site (Hayes et al. 1987). Neither of these two sites has thus far been implicated in NGF-mediated c-*fos* induction.

A host of agents which stimulate various second messenger pathways can induce c-*fos* in PC12 cells. One comparative analysis (Kruijer et al. 1985) of c-*fos* induction by NGF, EGF, PMA, K^+ depolarization, and cAMP elevation pointed out one common denominator among these agents' actions, the activation of C-kinase (McTigue et al. 1985; Cremins et al. 1986), and suggested that C-kinase may in part mediate c-*fos* induction by NGF. Consistent with this suggestion are the findings indicating that at least one NGF response element in the c-*fos* gene is also a PMA response element (SRE-I, see above).

Morgan and Curran (1986) examined the role of Ca^{++} in c-*fos* induction. They found that the K^+ depolarization induction of c-*fos* required the presence of extracellular Ca^{++} and was blocked by dihydropyridines, suggesting that the induction was mediated by Ca^{++} entry through voltage sensitive Ca^{++} channels. A role for calmodulin was further suggested since each of two calmodulin inhibitors, chlopromazine and trifluoperazine, blocked c-*fos* induction by depolarization. These drugs did not affect c-*fos* induction by PMA or by NGF, suggesting that calmodulin was not involved in c-*fos* induction by these agents. Based on these and other results, a model for c-*fos* induction was proposed, (Morgan and Curran 1986) involving protein kinase-mediated phosphorylation of a transcription activating protein (TAP). This type of protein kinase pathway was proposed to mediate c-*fos* induction by agents such as NGF and FGF.

A recent study from this laboratory (Armstrong and Halegoua, unpublished) has used a different approach to examine the pathways for c-*fos* induction by NGF. As discussed above, c-*fos* induction is transient, with transcription as well as RNA levels returning to near baseline within 2 h after the initial inducing stimulus. As with the human monocyte-like U937 cell line (Shibanuma et al. 1987), PC12 cells are refractory to reinduction of c-*fos* for up to 24 h depending upon the pathway used for the initial induction (Armstrong and Halegoua, unpublished). In PC12 cells, if c-*fos* is first induced by PMA (via C-kinase) or by depolarization (via Ca^{++}), reinduction by either of these pathways is not possible during the refractory period. However, in the same cells in which c-*fos* is initially induced with PMA or depolarization, reinduction of c-*fos* is possibly by cAMP elevation (via A-kinase). Initial induction of c-*fos* by cAMP likewise does not block reinduction by PMA or depolarization. Thus, two induction groups are defined by these experiments, one involving either C-kinase or Ca^{++}, and one involving cAMP. Using this same experimental paradigm to delineate the pathways involved in c-*fos* induction by NGF yielded unexpected results. Although reinduction of c-*fos* by NGF was blocked by prior treatment with NGF, c-*fos* was reinduced by NGF if cAMP was used as the conditioning agent. This result suggests that the cAMP pathway does not play a significant role in NGF-mediated c-*fos* induction. Initial induction of c-*fos* by PMA or depolarization significantly but only partly blocked (by about two-thirds) reinduction by

NGF, permitting about one-third the normal level of c-*fos* induction (about 15-fold) to occur, even in the A-kinase deficient mutant (A26-1B2). These results suggest that, although NGF shares a pathway for c-*fos* induction with C-kinase and/or Ca^{++}, another pathway for c-*fos* induction by NGF exists which is not mediated through C-kinase, Ca^{++}, or cAMP. How the two pathways revealed by the c-*fos* reinduction scheme described above relate to the NGF response elements (SRE-I and SRE-II) in the c-*fos* gene remains to be determined. The identity of the novel pathway for c-*fos* induction by NGF and its relationship to the *src* and *ras* components of NGF action may shed further light on the signal transduction mechanism for NGF-induced changes in gene expression.

Besides c-*fos*, several other primary response genes are induced by NGF. Two cDNA clones, designated NGFI-A (Milbrandt 1987) and NGFI-B (Milbrandt 1988), were isolated by differential screening of a PC12 library generated from PC12 cells treated with both NGF and cycloheximide for 3 h. *NGFI-A* and *-B* fall into the category of primary response genes based on the characteristics of their induction by NGF. They are induced in a time course similar to that of c-*fos*, and, like c-*fos*, *NGFI-A* is induced by PMA and by the Ca^{++} ionophore A23187. *NGFI-A* shows overlapping but also distinct characteristics with c-*fos* in the reinduction paradigm discussed above (Armstrong and Halegoua, unpublished).

Among five cDNA clones identified as representing primary response genes for mitogen (PMA)-stimulated fibroblasts (Lim et al. 1987), four were found to represent primary response gene transcripts in PC12 cells induced by NGF, EGF, PMA, or K^{+}-mediated depolarization (Kujubu et al. 1987). Interestingly, the fifth gene was not induced by any of these treatments of PC12 cells and thus probably represents a cell type or program (such as mitogenesis) specific transcript. This latter result raises the interesting possibility that there may exist primary response genes specific for the neuronal differentiation pathway (Kujubu et al. 1987). In addition to the five primary response genes described above, a host of other primary response gene transcripts have been similarly identified and cloned from fibroblasts (Cochran et al. 1983; Lau and Nathans 1985), some of which may be expressed in PC12 cells.

Several primary response genes have been identified on the basis of their immunological similarity to c-*fos* (Franza et al. 1987, 1988). These fos related antigens (fra) are induced with a rapid time course similar to c-*fos*, although the decay of fra mRNA can be less rapid. At least a dozen different fras have been identified suggesting the existence of a *fos* gene family. Another immediate early gene induced by NGF is the proto-oncogene c-*jun* (Wu et al. 1989). Again, the time course is rapid, although the level of induction is not as dramatic. A closely related gene, *jun-B*, is also induced by NGF (Bartel et al. 1989). In this case the kinetics of induction are somewhat slower than that observed for c-*jun* or c-*fos*.

The rapid kinetics and magnitude of primary response gene induction has been intriguing with respect to the potential role these gene products play in NGF action. In every case where functionality has been addressed, the result has pointed to their role as transcriptional trans-activators. The direct involvement of the c-*fos* gene product in the transcriptional activation of the collagenase gene (Schonthal et al. 1988) and of the preproenkephalin gene (Sonnenberg et al. 1989) has been demonstrated. Whether fos is acting as a component of the fos/jun containing AP1 complex in these cases has yet to be demonstrated. Although a requirement for c-*fos* expression in NGF-stimulated differentiation of PC12 cells has not been established, injection of specific anti-fos antibodies or expression of c-*fos* antisense RNA in fibroblasts inhibits the ability of serum to stimulate DNA synthesis and cell proliferation (Holt et al. 1986; Nishikura and Murray 1987; Riabowol et al. 1988). The NGF-stimulated primary response genes, *NGFI-A* and *NGFI-B*, have been characterized in enough detail as to suggest a function in transcriptional trans-activation. *NGFI-A* contains several sequence elements which are highly homologous to the DNA binding, zinc finger regions of many transcriptional regulatory proteins (Milbrandt 1987). The deduced amino acid sequence of *NGFI-B* reveals striking homologies to several members of the steroid hormone receptor family (Milbrandt 1988). There are two separate regions of homology, one encompasses the ligand binding domain and the other encompasses the DNA binding region of the receptor. These homologies suggest that perhaps the *NGFI-B* gene product may bind a steroid ligand which results in the activation of DNA binding and transcriptional trans-activation. This means of gene trans-activation is obviously different than those described above in that it would be dependent upon another hormone which may come from an exogenous source in vivo. Thus, NGF in this context would be a permissive agent for the action of other hormones during neuronal differentiation. The identification of the putative steroid ligand and the site of its synthesis will shed light on this intriguing question.

Primary response gene products have been suggested to act as "nuclear third messengers" in the response of cells to mitogenic, differentiation, or other long-term modulating stimuli (Morgan and Curran 1989), as outlined in Fig. 1. In this model, the primary response genes may provide the link between rapid post-translational events and long-term changes in gene expression. Within this context, specificity of responses could be provided in part by the precise array of TAPs initially expressed in the cell and the different combinatorial effects of primary response gene products (which may act as heterodimers) on subsequent gene expression. To date, a major aspect of this model remains to be tested, that primary response gene products cause changes in subsequent gene expression which lead to the changed phenotype (i.e., changes which play a causal role in NGF actions).

4.2 Other Genes

Subsequent gene expression changes, which may in many cases depend upon prior induction of primary response genes, generally depend upon ongoing protein synthesis and may or may not be of a transient nature. With only a fraction thus far identified, it is already apparent that subsequent changes in gene expression are quite varied in magnitudes of induction or repression and occur over very different time spans (see Fig. 4). Similar in one respect to *NGFI-B*, several "later" genes have been identified whose products may play a modulatory role in NGF action. One of these genes, *PC4* (TIRONE and SHOOTER 1989), is induced both quickly by NGF in PC12 cells, its RNA reaching a maximal level within 2 h, and transiently, with PC4 RNA gradually decreasing to baseline within 18 h of NGF treatment. Sequence analysis of the PC4 cDNA indicated that the predicted PC4 protein is partly homologous to the mouse putative β-interferon (IFN) protein and is highly homologous to rat τ-IFN, which has recently been shown to facilitate NGF action in PC12 cells (IMPROTA et al. 1988). Interestingly, PC4 was found to be expressed in the developing neural tube at times when neuroblasts are dividing and post-mitotic neurons are differentiating. The effects of PC4 protein on PC12 cell differentiation and its role in NGF action remains to be determined. One cDNA clone obtained from differential screening of cDNA libraries from naive vs long term NGF-treated PC12 cells, clone 63, was found to encode an mRNA which is maximally induced within 1 day of NGF treatment (LEONARD et al. 1987). The cDNA sequence is homologous to calcyclin (THOMPSON and ZIFF 1989), a growth regulated Ca^{++} binding protein found in a variety of cell types (FERRARI et al. 1987). Two other cDNA sequences, 42A and 42C (MASIAKOWSKI and SHOOTER 1988), represent mRNAs which maximally accumulate in PC12 cells after 24 h of NGF treatment and remain high for at least 7 days of treatment. The encoded proteins are homologous in sequence to each other and to the S-100 family of Ca^{++} binding proteins. The various Ca^{++} binding proteins may potentially play a role in the regulation of Ca^{++} or in the mediation of Ca^{++} effects during NGF-induced differentiation. Another NGF-induced protein, smg-25a, is a novel GTP binding protein which has recently been purified and the cDNA cloned from rat brain (SANO et al. 1989). This protein belongs to the family of ras-like G-proteins and is expressed preferentially in brain and in the adrenal medulla. In PC12 cells smg-25a expression is stimulated by NGF treatment to maximal levels within 1 day and levels remain high for at least 7 days. The precise function of this protein remains to be determined.

Since neurite outgrowth is a hallmark neuronal property, numerous studies of gene expression have focused on those genes whose expression is correlated with this event. Prominent among such genes are those encoding cytoskeletal filament proteins. The intermediate types filaments of PC12 cells consist of several different proteins which are regulated by NGF. As discussed above, analysis of protein levels and synthetic rates have shown that NGF increases the synthesis of peripherin and of neurofilament subunit proteins. A peripherin cDNA clone was originally identified as a clone (clone 73) corresponding to an mRNA which was maximally induced in PC12 cells after 2 days of NGF treatment (ALETTA et al. 1988b; LEONARD et al. 1988). Sequence analysis of clone 73 (LEONARD et al. 1988) and the gene encoding it (THOMPSON and ZIFF 1989) revealed it to be a type III intermediate filament protein (vimentin-like), and in situ hybridization analysis revealed a neuronal distribution limited to certain peripheral and central tissues (PARYSEK and GOLDMAN 1987; LEONARD et al. 1988). The potential significance of this new neuronal intermediate filament protein is intriguing. However, given our general lack of understanding of the role of intermediate filaments, the assignment of any physiological significance can only be speculative. As discussed above, the phosphorylation state as well as the synthesis of the three neurofilament subunit proteins are differentially regulated by NGF in PC12 cells. Differential regulation of mRNAs coding for the neurofilament proteins has also been observed. Both NF-L and NF-M mRNAs are increased within 7 days of NGF treatment. Increased NF-L mRNA was accompanied by increased transcription of the gene (Dickson et al. 1986; Lindenbaum et al. 1988), while the increased levels of NF-M mRNA were due to both increased gene transcription and metabolic stability (Lindenbaum et al. 1988). With NF-H, although the protein content was increased, the mRNA content, transcription, and stability were not found to be affected by NGF, suggesting control of the level of translation or of protein turnover, perhaps resulting from increased phosphorylation (Goldstein et al. 1987; Lindenbaum et al. 1987).

Among the cytoskeletal proteins whose synthetic rates are increased by NGF are those involved in the assembly of microtubules. Production of the structural components of microtubules, α- and β-tubulin, increases by about twofold during NGF-induced neurite outgrowth (Drubin et al. 1985; Black et al. 1986) due to an increased synthesis of the protein (Drubin et al. 1988)). Increased tubulin production can be accounted for by the twofold increase in its mRNA (Fernyhough and Ishii, 1987; Drubin et al. 1988). The expression of MAPs has also been investigated extensively. MAP 1.2 has been found to be increased during NGF-induced neuritogenesis (Greene et al. 1983; Drubin et al. 1985). The level of induction, initially complicated by concomitant changes in its phosphorylation state, has been determined to be about threefold (Aletta et al. 1988a) and can be explained by increased synthesis of the protein (Drubin et al. 1988) and the mRNA (Lewis et al. 1986; Aletta et al. 1988a). A relatively minor MAP species in PC12 cells, MAP 2, detected using specific antibodies, has also been found to be induced about twofold by long-term NGF treatment (Black et al. 1986; Brugg and Matus 1988). MAPs 3 and 5 have been found to be more dramatically increased, by 10-fold and 15-fold, respectively, after 10 days of NGF treatment (Brugg and Matus 1988). The regulation of MAPs 2, 3, and 5 have not yet been examined with respect to protein synthesis or mRNA levels. The "tau" MAPs protein levels, biosynthesis, and mRNA levels are all induced by NGF during neurite growth (Drubin et al. 1985, 1988), the potential significance of which will be discussed in Sect. 5.

Another group of NGF-induced proteins which may be important for neuritogenesis includes the growth associated protein GAP-43 (also termed B-50; see Skene 1989), Thy-1 glycoprotein (Richter-Landsberg et al. 1985; Dickson et al. 1986; Doherty and Walsh 1987), the so-called neural cell adhesion molecule N-CAM (Prentice et al. 1987), the adhesion-stimulating β-amyloid precursor (Schubert et al. 1989), and the protease transin (Machida et al. 1989). For each of these proteins both protein and mRNA levels have been shown to be increased by NGF in PC12 cells. Synthesis of thy-1 is also increased by FGF

(RYDEL and GREENE 1987). For both thy-1 (DICKSON et al. 1986) and transin the induction involves transcriptional regulation, while for GAP-43 the regulation has been suggested to be primarily post-transcriptional (FEDEROFF et al. 1988). By contrast, mRNA levels for NILE were found to remain unchanged after NGF treatment perhaps suggesting translational control (SAJOVIC et al. 1987). The induction of N-CAM and possibly the β-amyloid precursor involves differential splicing of primary gene transcripts resulting in the expression of different mRNAs and N-CAM forms. The potential significance of these gene expression changes to neurite growth will be discussed in Sect. 5.

A physiologically important molecule which is induced by NGF is the voltage dependent Na^+ channel. As discussed above, increases in Na^+ channel activity result in the establishment of a Na^+ based action potential within 4–7 days of NGF treatment. A further examination of NGF induction of Na^+ channel subtypes in PC12 cells was carried out using cDNA clones for type I and type II Na^+ channels obtained from rat brain (COOPERMAN et al. 1987). Using S1 nuclease protection assays, MANDEL et al. (1988) identified mRNAs corresponding to type II channels but not type I. The type II mRNA levels were induced up to eight fold by NGF treatment of the cells over 7 days, concomitant with the increase in voltage dependent Na^+ current. Thus, it appears likely that the increased Na^+ current seen in response to NGF is contributed to at least in part by the type II gene expression. It is not yet known whether additional Na^+ channel subtypes are also regulated by NGF, however, this possibility is strengthened by the appearance of TTX-insensitive Na^+ channels after NGF treatment (RUDY et al. 1987). Another physiologically relevant activity regulated by NGF is that of TH. In addition to the regulation of this enzyme by phosphorylation, as discussed above, long-term increased production of the enzyme has been seen in response to NGF in superior cervical ganglia (THOENEN et al. 1971; MAX et al. 1978). In PC12 cells the regulation appears to be more complex. PC12 cells contain a high basal level of TH activity, which decreases after long-term treatment with NGF (GREENE and TISCHLER 1976). Examination of the mRNA levels for TH by Northern blot analysis using cloned TH cDNA indicated that the mRNA is initially increased twofold within 4 h of NGF treatment and then gradually declines over the following week (LEONARD et al. 1987). However, using a subclone of PC12 cells, PC12h (HATANAKA 1981; HATANAKA and ARIMATSU 1984), or an independent clonal isolate from the original pheochromocytoma, PCG2, which contained low basal levels of TH expression (GOODMAN and HERSCHMAN 1978), NGF was found over a long time to induce TH activity and protein levels, similar to what was seen in superior cervical ganglion cultures.

Several additional cDNA clones have been isolated on the basis of NGF inducibility in PC12 cells, using differential screening. The represented mRNAs show varied time courses and magnitudes of induction by NGF. In the case of one such clone, VGF-8 (LEVI et al. 1985), an 80-fold transcriptional induction of the mRNA occurs within 9 h of NGF treatment. The 76 kDa protein sequence has been deduced from the cDNA, and antibodies have been generated against fusion products expressed in *E. coli* (Possenti et al. 1989). Using these antibodies, the protein has been found to be contained in and released from secretory vesicles. Whether this protein plays a role in neuronal function and/or the differentiation process remains to be determined. Another cDNA clone, SCG10, which was originally isolated by differential screening of rat sympathetic neurons and chromaffin cells, was found to represent an RNA which is induced within a 10 h NGF treatment of PC12 cells (Anderson and Axel 1985). This induction is mediated at least in part by transcriptional activation of the *SCG10* gene (Stein et al. 1988). One cDNA clone was found to be highly homologous to thymosin β-4 (Leonard et al. 1987), a protein with a wide tissue distribution. The mRNA for this protein is induced up to tenfold, beginning after 12 h of NGF treatment. Although previously characterized, the function of this protein remains to be determined. Several of the cDNA clones which have been isolated on the basis of their representing sequences for RNAs which are induced by NGF do not represent known genes or have as yet not been as well-characterized (Leonard et al. 1987; Cho et al. 1989). Characterization of these and other NGF-induced genes yet to be identified is expected to yield insights into the mechanism of NGF action as well as identify cellular components important for the neuronal phenotype.

Clearly, expression of a wide variety of genes are regulated during NGF-induced differentiation of PC12 cells. The enormous range of time courses for these events, which have been summarized in Fig. 4, suggests a variety of modes of gene regulation. Although the primary response genes appear to be directly regulated by second messenger activities rapidly activated in response to NGF, the regulation of later gene expression is much less clear. Transcriptional trans-activation by primary response genes undoubtedly plays a role in later gene regulation and evidence for this has been accumulating for some genes such as *transin* (Kerr et al. 1988) and *SCG10* (Stein et al. 1988). However, a number of genes regulated by NGF show differential regulation by second messengers. For example, different genes can be regulated either by C-kinase activation, cAMP elevation, Ca^{++}, or by subsets of these (Leonard et al. 1987; Cho et al. 1989). Elucidation of the pathways for the different gene inductions will undoubtedly unveil a complex interplay among second messenger systems, transcriptional trans-activators, and modulatory gene products which interact in precise ways within a specific time course in order to elicit the proper differentiation process. Evidence for these interactions has been emerging from studies of post-translational modification of regulated gene products. The c-*fos* gene, for example, is extensively post-translationally modified after induction by NGF treatment of PC12 cells (Curran and Morgan 1986). Similar findings have been reviewed above for other NGF-induced gene products such as MAP 1, peripherin, and TH. While such examples represent integration of post-translational modification and gene regulation in order to achieve a common goal (e.g. increased activity), they also suggest modulation of induced protein function over time in order to achieve a precise program of differentiation.

5 Neuritic Process Outgrowth

A dramatic illustration of NGF action in PC12 cells is the initiation as well as the directional elongation of neuritic processes, which requires the continual presence of the growth factor. Although extensively studied in a variety of neuronal culture systems (see for reviews SMITH 1988; MITCHISON and KIRSCHNER 1988), the molecular mechanism for neurite growth is largely unknown. We will concentrate on those PC12 cell components whose modulation by NGF may underlie neurite growth.

In a variety of structural and ultrastructural details, the progression of PC12 neurite growth closely resembles that of neurons (JACOBS and STEVENS 1986). NGF treatment of PC12 results in the formation of a neurite with longitudinal arrays of packed microtubules and some neurofilaments and containing typical microfilaments of the cortical membrane and cytoplasm. Characteristic growth cones are found at the neurite ends, although the filopodial structures are typically smaller than those found on cultured peripheral neurons. As the neurite grows, new membrane is added at the growth cone. Studies on neurite growth in culture have revealed characteristics which have led to two generally opposing models for neurite growth. In the first model, both adhesion of the growth cone to the growth substrate and the tension generated along the neurite direct the addition of membrane at the growth cone and determine the directionality of growth. This model has been supported by a variety of data (for review see SMITH 1988; MITCHISON and KIRSCHNER 1988). Filopodia, with ordered arrays of actin and associated myelin containing microfilaments, have also been found to be sites of focal contact to the substratum (LETOURNEAU 1979, 1981) suggesting that filopodial structures are the source of adhesion and tension. Later experiments, however, indicated that neurites could grow (although abnormally) in the absence of filopodial movement which was inhibited by cytochalasin (which blocks microfilament function) (MARSH and LETOURNEAU 1984) or in the absence of a suitable adhesive substrate (STRASSMAN et al. 1973). More recently, high resolution, video-enhanced, contrast-differential, interference contrast microscopy was used to examine growth cone movements (GOLDBERG and BURMEISTER 1986; ALETTA and GREENE 1988). Based on these observations, an alternative model for neurite growth was proposed in which membrane addition at the growth cone occurs in the form of motile lamellipodia followed by axoplasmic flow in conjunction with modification of substrate attachment by various areas of the growth cone. Neurite growth was thus proposed to occur by cytoplasmic "pushing" from the growth cone rather than by growth cone "pulling". It would thus appear that both type of events (pushing and pulling) occur at the growth cone and must somehow be coordinated to control neurite growth under different environmental conditions (i.e. varying substrata).

NGF has been shown to dramatically influence neurite outgrowth and growth cone movements from PC12 and primary neurons. NGF elicits de novo neurite outgrowth from PC12 cells; the elongating neurite requires continual

NGF action; and its growth may follow an NGF concentration gradient. These effects of NGF, as with the many other effects discussed above, appear to follow the same overall model for NGF action shown in Fig. 1, in which a combination of rapid post-translational events combined with longer term gene expression changes lead to changes in both activity and levels of the essential components for the response. This is evident from the long lag time and transcriptional dependence of neurite initiation caused by NGF (see above), the transcriptional independence of neurite regeneration (also see above), and the rapid effects of NGF on growth cone structure.

5.1 Adhesion Molecules and the Cell Surface

The adhesive properties of PC12 cells and the effects of NGF on PC12 adhesion have been studied under a variety of culture conditions. PC12 cells will adhere to different substrates including tissue culture plastic, polylysine, laminin, fibronectin, and collagen. The attachment of PC12 cells to each of these substrates appears to be mediated by members of the integrin family of matrix receptors. For example, anti-integrin antibodies, or peptides corresponding to fibronectin binding sites to integrins, are able to partially block PC12 adhesion and profoundly block neurite outgrowth on these substrates (SCHWARZ et al. 1989; TURNER et al. 1989). Several studies have shown that NGF rapidly stimulates the substrate adhesion of PC12 cells (SCHUBERT and WHITLOCK 1977; SCHUBERT et al. 1978; CHANDLER and HERSCHMAN 1980; FUJII et al. 1982; AKESON and WARREN 1986) and changes surface properties of the cells (NAOI et al. 1988), suggesting that these changes are important to NGF-induced neurite growth (SCHUBERT et al. 1978). However, although NGF stimulates the adhesion of PC12 to the substrate within minutes, on a longer time scale of days, the overall adhesion of cells with neurites is diminished. An examination of PC12 adhesion sites on polylysine and collagen clarifies this apparent discrepancy. When the adhesion of PC12 cells to the substrate was examined by interference reflection microscopy, focal adhesions were found to be limited to the cell periphery, while only close contacts were found at internal regions (HALEGOUA 1987). Within hours of NGF treatment of these cells, a redistribution and pronouncement of focal contact sites was observed at protruding edges of the cell periphery (HALEGOUA 1987). With longer term NGF treatment, focal contacts remained with the growth cone and were rarely seen along the neurite or cell body (HALEGOUA 1987). Thus, as is seen in primary neuron cultures (LETOURNEAU 1979, 1981), growth cones are the major sites of substrate adhesion for neurite containing PC12 cells. It is interesting that the integrins are found throughout the cell membrane, suggesting that other membrane and/or cytoskeletal components are additionally needed for directing focal contact formation (see below).

Cell surface and secreted components of PC12 cells have been identified whose synthesis is stimulated by NGF in a time course consistent with their proposed involvement in neurite growth. One such substance is a heparan sulfate proteoglycan which promotes neurite growth in PC12 cells as well as in various primary cultures of peripheral neurons (LANDER et al. 1983). The proteoglycan is active when complexed with laminin, and antibodies against the complex can block NGF directed neurite growth (CHIU et al. 1986). It has not been determined whether NGF stimulates the synthesis of the proteoglycan although changes in overall heparan sulfate glycosaminoglycans of PC12 cells resulting from long-term NGF treatment have been reported (MARGOLIS et al. 1987). The PC12 cell surface glycoprotein, NILE, synthesis and glycosylation is regulated by NGF treatment of PC12 cells. This neuron specific protein has also been independently determined by several groups to be involved in growth cone-axon and axon-axon adhesions (also termed L1, NG-CAM, G4, 8D9, ASCS4, 69A1) (JESSELL 1988). Antibodies against NILE block neurite fasciculation (STALLCUP and BEASLEY 1985) and axon elongation along axons but not along extracellular matrix (CHANG et al. 1987), suggesting a role for this protein in both fasciculation and elongation. The protein can also provide a suitable substrate for neurite growth (LAGENAUR and LEMMON 1987), suggesting a possible role in adhesion. Another glycoprotein whose synthesis is stimulated by NGF is thy 1.1 (see above). That it may be involved in neurite growth is suggested by its time course of synthesis and its ability to provide a suitable substrate for neurite growth (MAKANTHAPPA and PATTERSON 1989). It is thus possible that thy 1.1 plays a general role in the adhesive process. A neural cell adhesion molecule, N-CAM, has also been detected in PC12 cells and its synthesis has also been found to be increased by NGF treatment (see above). While this protein may have a variety of functions in cell adhesion during development (RUTISHAUSER and JESSELL 1988), it appears to be involved in neurite fasciculation as opposed to neurite outgrowth (CHANG et al. 1987). It should be kept in mind that the increased synthesis of the above proteins in response to NGF may reflect a result of the NGF promotion of neuronal differentiation which would include a variety of adhesive and other functions unrelated to the mechanism of NGF actions.

5.2 Roles of the Cytoskeleton

As mentioned above, the cytoskeleton of the neurite is an ordered array consisting of linearly packed microtubules and of actin microfilaments packed predominantly as a cortical shell beneath the membrane. At the growth cone, the assemblies of these structures are much more dynamic. Actin or tubulin are continually assembled into short lived polymers and clearly play important roles in determining the growth cone's protrusive activities (recently reviewed by MITCHISON and KIRSCHNER 1988). Neurofilaments can be occasionally found along the neurite, but the dynamics of their assembly and relative absence in early stages of growing neurites are inconsistent with their involvement in neurite growth, in spite of the NGF stimulated synthesis of neurofilament and other intermediate filament subunit proteins (see above). However, these structures may play a role in maintaining the structural integrity of the neurite.

While actin filament assembly and movements are likely to be playing a major role in protrusive activities at the growth cone, the precise ways in which this is accomplished are not presently understood. One attractive model for these movements involves monomeric actin and the retrograde movements of actin filaments presumably via interaction with type I myosin (MITCHISON and KIRSCHNER 1988; SMITH 1988). Clearly, the tension generated at the growth cone would involve coordination of these processes with those involved in the establishment and control of focal adhesion sites (i.e., sites of integrin attachment to substrate molecules). The linkage between the integrin receptors and actin filaments is mediated directly through proteins such as talin and accessory molecules such as vinculin (BURRIDGE et al. 1988). Since establishment of these assemblies may be controlled by post-translational events, such as phosphorylation and proteolysis, they represent a potential site for control of neurite growth. It is interesting in this regard that NGF treatment has been found to result in vinculin phosphorylation concomitant with the redistribution of vinculin foci and focal contact sites preceding neurite outgrowth in PC12 cells (HALEGOUA 1987). This study suggested that NGF-stimulated phosphorylation of proteins, such as vinculin, might help establish tension directed neurite outgrowth via changes in focal adhesion sites. A variety of other potential control processes have also been discussed (MITCHISON and KIRSCHNER 1988).

Microtubules of the growing neurite are also in a dynamic state of assembly/disassembly and the control of these processes is essential to neurite outgrowth. Although microfilament interactions at focal contacts may control orientation of the growth cone, an initial phase of neurite growth, microtubule polymerization and stabilization controls stabilization of the neurite during growth and polarization of the growing process (MITCHISON and KIRSCHNER 1988). Microtubule assembly in the growth cone, unlike that along the neurite, is in a state of rapid flux, alternating between assembly and disassembly (OKABE and HIROKAWA 1988). The stabilization and bundling of polymerized microtubules is essential to neurite growth, as is evident by the block of neurite growth caused by microtubule-destabilizing drugs such as colchicine. A variety of MAPs such as tau, MAP 1.1 and 2, and several chartins, have been described which may in part control these functions. Regulation of both the abundance and post-translational processing of MAPs in response to NGF (see above and BLACK and KEYSER 1987) has been proposed to result in the stabilization of microtubules essential to neurite growth. Thus NGF may influence microtubule assembly both long-term (by increased gene expression of tubulin and MAPs) and short-term (by post-translational modification of MAPs) leading to the establishment and elongation of neurites.

5.3 Regulatory Molecules at the Growth Cone

Our discussion of neurite growth thus far has been limited to the structural components of the growth cone. However, we have alluded to the fact that NGF has various rapid effects at the growth cone and may control structural features of the growth cone via post-translational modifications. This requires that NGF receptors be located at the growth cone. In fact, NGF receptors are preferentially located at neuritic growth cones (Levi et al. 1980) and NGF applied to the neurite can influence events at the cell soma. Thus, it would appear likely that the NGF transducing molecules might also be preferentially localized there. Consistent with this, the c-src tyrosine kinase has been found to be concentrated at the growth cone (Sobue and Kanda 1988). Whether c-src plays any direct role in controlling neurite growth or is simply a signal transducer for NGF will await identification of the c-src tyrosine kinase substrates. However, it is of interest that the location of v-src protein, in PC12 cells transfected and induced to differentiate by the v-*src* gene, appears to be localized at growth cone focal adhesion plaques (J. Halegoua and S. Halegoua, unpublished work). Determination of the roles of other potential NGF signal transducing molecules such as c-ras should shed further light on the mechanics of growth cone control.

Another regulatory molecule which has received much attention is GAP-43. Originally discovered as a protein whose synthesis is well-correlated with axonal growth and regeneration (Skene 1989), this protein is found in neuritic processes and is prominent in growth cones of NGF treated PC12 cells (Van Hooff et al. 1989). Treatment of PC12 cells with NGF results within 48 h in the redistribution of GAP-43 from cytoplasmic membranous structures to a plasma membrane location especially associated with filopodial and other protrusions. GAP-43 expression is also increased by NGF by about 3.5-fold (see above). That this protein may play an important role in mediating neurite growth is indicated by its ability to elicit filopodial processes in GAP-43 gene transfected Chinese hamster ovary (CHO) cells (Zuber et al. 1989). GAP-43 appears to be intimately involved in second messenger regulation. It is a prominent substrate for C-kinase in growth cone membranes (Van Hooff et al. 1988) and is phosphorylated in response to NGF (K. Meiri, personal communication). It has thus been implicated in feedback regulation of phosphatidylinositol turnover (Van Hooff et al. 1988). GAP-43 has recently been found to be a calmodulin binding protein which binds calmodulin in low Ca^{++} and releases it in high Ca^{++}. These unusual calmodulin binding properties suggests that it may regulate calmodulin-Ca^{++} interactions (Andreasen et al. 1983; Cimler et al. 1987; Skene 1989). The precise role of GAP-43 in growth cone function, its relationship to other structural and regulatory molecules in the growth cone, and the potential importance of its regulation by NGF should provide important insights into understanding the regulation of neurite growth.

6 Conclusions

Studies carried out over the last decade have established a molecular framework for NGF action in PC12 cells. The overview for this framework, depicted in Fig. 1, has received much experimental support. A variety of second messenger pathways stimulated in response to NGF have been identified. A number of post-translational modifications have been described and characterized. In some instances, specific protein kinase activations and protein phosphorylations have been shown or been suggested to mediate acute physiologic actions of NGF such as increased neurotransmitter synthesis. Rapid events appear to be continually necessary for the ongoing actions of NGF throughout the differentiation process involving also neurite outgrowth. Evidence has also been accumulating for a role of protein phosphorylation in mediating changes in gene expression. Such changes are essential for long-term NGF actions such as the establishment of the action potential mechanism and the generation of neurites. Insights into the signal transduction process have also been gained by the observation that proto-oncogenes appear to play essential roles in mediating major actions of NGF. It appears likely that the general model for NGF action can be extrapolated to the actions of FGF and other neuronal growth factors.

Although a general framework for NGF actions has been established, this framework is lacking in molecular detail and major issues remain to be resolved. For example, the *src* and *ras* proto-oncogene products have been identified as components of the NGF signal transduction machinery, but their precise relationships to the NGF receptor and to each other and to the second messenger pathways to which they may be coupled remain to be determined. The interactions of the various second messenger pathways themselves have only sketchily been mapped. Evidence for specific molecular events underlying second messenger-mediated physiological changes is now only fragmentary and identification of the essential genes for the NGF-differentiation process and their mode of regulation has just begun. Novel signal transduction events and additional second messenger pathway interactions which underlie NGF action almost certainly exist and await discovery.

Studies on NGF action will continue to feed into and in turn be fed by advances in other related fields of study. Studies of the control of cellular metabolism, cell shape determination, and the control of normal and malignant cell growth all involve coordinate regulation of genes and second messenger molecules characteristic of the NGF response. In this regard, it is interesting that the overall framework for NGF-induced differentiation strikingly resembles emerging models for neuronal plasticity (see for example Goelet et al. 1986). Studies of the program of NGF-induced neuronal differentiation may thus provide more general insights into growth factor and neurotransmitter modulation of functions in adult neurons. Advances in our understanding of these signal transduction processes will undoubtedly continue to provide an arena for a new set of exciting studies over the next decade.

Acknowledgements. The authors thank Dr. Gail Mandel for advice, encouragement, and critical reading of the manuscript; Dr. Paul Brehm for help in the design of the figures; and Dr. Joel Levine for helpful suggestions for the section of neuritic process outgrowth. This laboratory has been supported by a grant from the National Institutes of Health (NINDS).

References

Abdel-Latif AA (1986) Calcium-mobilizing receptors, polyphosphoinositides, and the generation of second messengers. Pharmacol Rev 38: 227–272

Akeson R, Warren SL (1986) PC12 adhesion and neurite formation on selected substrates are inhibited by some glycosaminoglycans and a fibronectin-derived tetrapeptide. Exp Cell Res 162: 347–362

Albert KA, Helmer-Matyjek E, Nairn AC, Müller TH, Haycock JW, Greene LA, Goldstein M, Greengard P (1984) Calcium/phospholipid-dependent protein kinase (protein kinase C) phosphorylates and activates tyrosine hydroxylase. Proc Natl Acad Sci USA 81: 7713–7717

Alemà S, Casalbore P, Agostini E, Tatò F (1985) Differentiation of PC12 phaeochromocytoma cells induced by v-*src* oncogene. Nature 316: 557–559

Aletta JM, Greene LA (1988) Growth cone configuration and advance: a time-lapse study using video-enhanced differential interference contrast microscopy. J Neurosci 8: 1425–1435

Aletta JM, Lewis SA, Cowan NJ, Greene LA (1988a) Nerve growth factor regulates both the phosphorylation and steady-state levels of microtubule-associated protein 1.2 (MAP 1.2). J Cell Biol 106: 1573–1581

Aletta JM, Angeletti R, Liem RK, Purcell C, Shelanski ML, Greene LA (1988b) Relationship between the nerve growth factor-regulated clone 73 gene product and the 58-kilodalton neuronal intermediate filament protein (peripherin). J Neurochem 51: 1317–1320

Aletta JM, Shelanski ML, Greene LA (1989) Phosphorylation of the peripherin 58-kDa neuronal intermediate filament protein. Regulation by nerve growth factor and other agents. J Biol Chem 264: 4619–4627

Allen JM, Tischler AS, Lee YC, Bloom SR (1984) Neuropeptide Y (NPY) in PC12 phaeochromocytoma cultures: responses to dexamethasone and nerve growth factor. Neurosci Lett 46: 291–296

Anderson DJ, Axel R (1985) Molecular probes for the development and plasticity of neural crest derivatives. Cell 42: 649–662

Andreasen TJ, Luetje CW, Heideman W, Storm DR (1983) Purification of a novel calmodulin binding protein from bovine cerebral cortex membranes. Biochemistry 22: 4615–4618

Andres RY, Jeng I, Bradshaw RA (1977) Nerve growth factor receptors: identificition of distinct classes in plasma membranes and nuclei of embryonic dorsal root neurons. Proc Natl Acad Sci USA 74: 2785–2789

Bar-Sagi D, Feramisco JR (1985) Microinjection of the *ras* oncogene protein into PC12 cells induces morphological differentiation. Cell 42: 841–848

Bar-Sagi D, Feramisco JR (1986) Induction of membrane ruffling and fluid-phase pinocytosis in quiescent fibroblasts by *ras* proteins. Science 233: 1061–1068

Barbacid M (1987) *ras* Genes. Annu Rev Biochem 56: 779–827

Bartel DP, Sheng M, Lau LF, Greenberg ME (1989) Growth factors and membrane depolarization activate distinct programs of early response gene expression: dissociation of *fos* and *jun* induction. Genes Dev 3: 304–313

Biocca S, Cattaneo A, Calissano P (1984) Nerve growth factor inhibits the synthesis of a single-stranded DNA binding protein in pheochromocytoma cells (clone PC12). Proc Natl Acad Sci USA 81: 2080–2084

Black MM, Keyser P (1987) Acetylation of alpha-tubulin in cultured neurons and the induction of alpha-tubulin acetylation in PC12 cells by treatment with nerve growth factor. J Neurosci 7: 1833–1842

Black MM, Aletta JM, Greene LA (1986) Regulation of microtubule composition and stability during nerve growth factor-promoted neurite outgrowth. J Cell Biol 103: 545–557

Blenis J, Erikson RL (1986) Regulation of protein kinase activities in PC12 pheochromocytoma cells. EMBO J 5: 3441–3447

Bothwell MA, Schechter AL, Vaughn KM (1980) Clonal variants of PC12 pheochromocytoma cells with altered response to nerve growth factor. Cell 21: 857–866

Brugg B, Matus A (1988) PC12 cells express juvenile microtubule-associated proteins during nerve growth factor-induced neurite outgrowth. J Cell Biol 107: 643–650

Burridge K, Fath K, Kelly T, Nuckolls G, Turner C (1988) Focal adhesions: transmembrane junctions between the extracellular matrix and the cytoskeleton. Annu Rev Cell Biol 4: 487–525

Burstein DE, Seeley PJ, Greene LA (1985) Lithium ion inhibits nerve growth factor-induced neurite outgrowth and phosphorylation of nerve growth factor-modulated microtubule-associated proteins. J Cell Biol 101: 862–870

Calissano P, Volonte C, Biocca S, Cattaneo A (1985) Synthesis and content of a DNA-binding protein with lactic dehydrogenase activity are reduced by nerve growth factor in the neoplastic cell line PC12. Exp Cell Res 161: 117–129

Campenot RB (1977) Local control of neurite development by nerve growth factor. Proc Natl Acad Sci USA 74: 4516–4519

Chan BL, Chao MV, Saltiel AR (1989) Nerve growth factor stimulates the hydrolysis of glycosylphosphatidylinositol in PC-12 cells: a mechanism of protein kinase C regulation. Proc Natl Acad Sci USA 86: 1756–1760

Chandler CE, Herschman HR (1980) Tumor promoter modulation of epidermal growth factor- and nerve growth factor-induced adhesion and growth factor binding of PC-12 pheochromocytoma cells. J Cell Physiol 105: 275–285

Chang S, Rathjen FG, Raper JA (1987) Extension of neurites on axons is impaired by antibodies against specific neural cell surface glycoproteins. J Cell Biol 104: 355–362

Chiu AY, Matthew WD, Patterson PH (1986) A monoclonal antibody that blocks the activity of a neurite regeneration-promoting factor: studies on the binding site and its localization in vivo. J Cell Biol 103: 1383–1398

Cho KO, Skarnes WC, Minsk B, Palmieri S, Jackson-Grusby L, Wagner JA (1989) Nerve growth factor regulates gene expression by several distinct mechanisms. Mol Cell Biol 9: 135–143

Cimler BM, Giebelhaus DH, Wakim BT, Storm DR, Moon RT (1987) Characterization of murine cDNAs encoding P-57, a neural-specific calmodulin-binding protein. J Biol Chem 262: 12158–12163

Cochran BH, Reffel AC, Stiles CD (1983) Molecular cloning of gene sequences regulated by platelet-derived growth factor. Cell 333: 939–947

Comb M, Birnberg NC, Seasholtz A, Herbert E, Goodman HM (1986) A cyclic AMP- and phorbol ester-inducible DNA element. Nature 323: 353–356

Contreras ML, Guroff G (1987) Calcium-dependent nerve growth factor-stimulated hydrolysis of phosphoinositides in PC12 cells. J Neurochem 48: 1466–1472

Cooperman SS, Grubman SA, Barchi RL, Goodman RH, Mandel G (1987) Modulation of sodium-channel mRNA levels in rat skeletal muscle. Proc Natl Acad Sci USA 84: 8721–8725

Cremins J, Wagner JA, Halegoua S (1986) Nerve growth factor action is mediated by cyclic AMP- and Ca^{+2}/phospholipid-dependent protein kinases. J Cell Biol 103: 887–893

Curran T (1988) The *fos* oncogene. In: Reddy EP, Skalka AM (eds) The oncogene handbook. Elsevier, Amsterdam, pp 307–325

Curran T, Morgan JI (1986) Barium modulates c-*fos* expression and post-translational modification. Proc Natl Acad Sci USA 83: 8521–8524

DeGeorge JJ, Walenga R, Carbonetto S (1988) Nerve growth factor rapidly stimulates arachidonate metabolism in PC12 cells: potential involvement in nerve fiber growth. J Neurosci Res 21: 323–332

Dichter MA, Tischler AS, Greene LA (1977) Nerve growth factor-induced increase in electrical excitability and acetylcholine sensitivity of a rat pheochromocytoma cell line. Nature 268: 501–504

Dickson G, Prentice H, Julien J-P, Ferrari G, Leon A, Walsh FS (1986) Nerve growth factor activates Thy-1 and neurofilament gene transcription in rat PC12 cells. EMBO J 5: 3449–3453

Doherty P, Walsh FS (1987) Control of Thy-1 glycoprotein expression in cultures of PC12 cells. J Neurochem 49: 610–616

Domenico Borasio G, John J, Wittinghofer A, Barde Y-A, Sendtner M, Heumann R (1989) *ras* p21 Protein promotes survival and fiber outgrowth of cultured embryonic neurons. Neuron 2: 1087–1096

Drubin D, Kobayashi S, Kellogg D, Kirschner M (1988) Regulation of microtubule protein levels during cellular morphogenesis in nerve growth factor-treated PC12 cells. J Cell Biol 106: 1583–1591

Drubin DG, Feinstein SC, Shooter EM, Kirschner MW (1985) Nerve growth factor-induced neurite outgrowth in PC12 cells involves the coordinate induction of microtubule assembly and assembly-promoting factors. J Cell Biol 101: 1799–1807

Ellis C, Moran M, McCormick F, Pawson T (1990) Phosphorylation of GAP and GAP-associated proteins by transforming and mitogenic tyrosine kinases. Nature 343: 377–381

End D, Hanson M, Hashimoto S, Guroff G (1982) Inhibition of the phosphorylation of a 100000 dalton soluble protein in whole cells and cell-free extracts of PC12 pheochromocytoma cells following treatment with nerve growth factor. J Biol Chem 257: 9223–9225

End D, Tolson N, Hashimoto S, Guroff G (1983) Nerve growth factor-induced decrease in the cell-free phosphorylation of a soluble protein in PC12 cells. J Biol Chem 258: 6549–6555

Federoff HJ, Grabczyk E, Fishman MC (1988) Dual regulation of GAP-43 gene expression by nerve growth factor and glucocorticoids. J Biol Chem 263: 19290–19295

Feig LA, Cooper GM (1988) Inhibition of NIH 3T3 cell proliferation by a mutant *ras* protein with preferential affinity for GDP. Mol Cell Biol 8: 3235–3243

Fernyhough P, Ishii DN (1987) Nerve growth factor modulates tubulin transcript levels in pheochromocytoma PC12 cells. Neurochem Res 12: 891–899

Ferrari G, Greene LA (1985) Does phospholipid methylation play a role in the primary mechanism of action of nerve growth factor. J Neurochem 45: 853–859

Ferrari S, Calabretta B, DeRiel JK, Battini R, Ghezzo F, Lauret E, Griffin C, Emanuel BS, Gurrieri F, Baserga R (1987) Structural and functional analysis of a growth-regulated gene, the human calcyclin. J Biol Chem 262: 8325–8332

Fisch TM, Prywes R, Simon MC, Roeder RG (1989) Multiple sequence elements in the c-*fos* promoter mediate induction by cAMP. Genes Dev 3: 198–211

Fleischman LF, Chahwala SB, Cantley L (1986) *ras*-Transformed cells: altered levels of phosphatidylinositol-4, 5-bisphosphate and catabolites. Science 231: 407–410

Franza BR Jr, Sambucetti LC, Cohen DR, Curran T (1987) Analysis of fos protein complexes and fos-related antigens by high-resolution two-dimensional gel electrophoresis. Oncogene 1: 213–221

Franza BR Jr, Rauscher FJ III, Josephs SF, Curran T (1988) The fos complex and fos-related antigens recognize sequence elements that contain AP-1 binding sites. Science 239: 1150–1153

Frazier WA, Ohlendorf CE, Boyd LF, Aloe L, Johnson EM, Ferrendelli JA, Bradshaw RA (1973) Mechanism of action of nerve growth factor and cyclic AMP on neurite outgrowth in embryonic chick sensory ganglia: demonstration of independent pathways of stimulation. Proc Natl Acad Sci USA 70: 2448–2452

Fujii DK, Massoglia SL, Savion N, Gospodarowicz D (1982) Neurite outgrowth and protein synthesis by PC12 cells as a function of substratum and nerve growth factor. J Neurosci 2: 1157–1175

Fujii M, Shalloway D, Verma IM (1989) Gene regulation by tyrosine kinases: src protein activates various promoters, including c-*fos*. Mol Cell Biol 9: 2493–2499

Garber SS, Hoshi T, Aldrich RW (1989) Regulation of ionic currents in pheochromocytoma cells by nerve growth factor and dexamethasone. J Neurosci 9: 3976–3987

Garrels JI, Schubert D (1979) Modulation of protein synthesis by nerve growth factor. J Biol Chem 254: 7978–7985

Gilman AG (1984) G proteins and dual control of adenylate cyclase. Cell 36: 577–579

Goelet P, Castellucci VF, Schacher S, Kandel ER (1986) The long and the short of long-term memory—a molecular framework. Nature 322: 419–422

Goldberg DJ, Burmeister DW (1986) Stages in axon formation: observations of growth of *Aplysia* axons in culture using video-enhanced contrast-differential interference contrast microscopy. J Cell Biol 103: 1921–1931

Goldstein ME, Sternberger NH, Sternberger LA (1987) Phosphorylation protects neurofilaments against proteolysis. J Neuroimmunol 14: 149–160

Goodman R, Herschman HR (1978) Nerve growth factor-mediated induction of tyrosine hydroxylase in a clonal pheochromocytoma cell line. Proc Natl Acad Sci USA 75: 4587–4590

Greenberg ME, Ziff EB (1984) Stimulation of 3T3 cells induces transcription of the c-*fos* proto-oncogene. Nature 311: 433–438

Greenberg ME, Greene LA, Ziff EB (1985) Nerve growth factor and epidermal growth factor induce rapid transient changes in proto-oncogene transcription in PC12 cells. J Biol Chem 260: 14101–14110

Greene LA, Rein G (1977) Synthesis, Storage and release of acetylcholine by a noradrenergic pheochromocytoma cell line. Nature 268: 349–351

Greene LA, Shooter EM (1980) The nerve growth factor: biochemistry, synthesis, and mechanism of action. Annu Rev Neurosci 3: 353–402

Greene LA, Tischler AS (1976) Establishment of a noradrenergic clonal line of rat adrenal pheochromocytoma cells which respond to nerve growth factor. Proc Natl Acad Sci USA 73: 2424–2428

Greene LA, Tischler AS (1982) PC12 pheochromocytoma cultures in neurobiological research. Adv Cell Neurobiol 3: 373–414

Greene LA, Burstein DE, Black MM (1982) The role of transcription-dependent priming in nerve growth factor promoted neurite outgrowth. Dev Biol 91: 305–316

Greene LA, Liem RK, Shelanski ML (1983) Regulation of a high molecular weight microtubule-associated protein in PC12 cells by nerve growth factor. J Cell Biol 96: 76–83

Guerrero I, Wong H, Pellicer A, Burstein DE (1986) Activated N-*ras* gene induces neuronal differentiation of PC12 rat pheochromocytoma cells. J Cell Physiol 129: 71–76

Guerrero I, Pellicer A, Burstein DE (1988) Dissociation of c-*fos* from ODC expression and neuronal differentiation in a PC12 subline stably transfected with an inducible N-*ras* oncogene. Biochem Biophys Res Commun 150: 1185–1192

Gunning PW, Landreth GE, Bothwell MA, Shooter EM (1981) Differential and synergistic actions of nerve growth factor and cyclic AMP in PC12 cells. J Cell Biol 89: 240–245

Hagag N, Halegoua S, Viola M (1986) Inhibition of growth factor-induced differentiation of PC12 cells by microinjection of antibody to *ras* p21. Nature 319: 680–682

Halegoua S (1987) Changes in the phosphorylation and distribution of vinculin during nerve growth factor induced neurite outgrowth. Dev Biol 121: 97–104

Halegoua S, Patrick J (1980) Nerve growth factor mediates phosphorylation of specific proteins. Cell 22: 571–581

Hall FL, Fernyhough P, Ishii DN, Vulliet PR (1988) Suppression of nerve growth factor-directed neurite outgrowth in PC12 cells by sphingosine, an inhibitor of protein kinase C. J Biol Chem 263: 4460–4466

Hama T, Guroff G (1985) Distribution of Nsp100 and Nsp100 kinase, a nerve growth factor-sensitive phosphorylation system, in rat tissues. J Neurochem 45: 1279–1287

Hama T, Huang KP, Guroff G (1986) Protein kinase C as a component of a nerve growth factor-sensitive phsophorylation system in PC12 cells. Proc Natl Acad Sci USA 83: 2353–2357

Hashimoto S (1988) K-252a, a potent protein kinase inhibitor, blocks nerve growth factor-induced neurite outgrowth and changes in the phosphorylation of proteins in PC12h cells. J Cell Biol 107: 1531–1539

Hashimoto S, Iwasaki C, Kuzuya H, Guroff G (1986) Regulation of nerve growth factor action on Nsp100 phosphorylation in PC12h cells by calcium. J Neurochem 46: 1599–1604

Hatanaka H (1981) Nerve growth factor-mediated stimulation of tyrosine hydroxylase activity in a clonal rat pheochromocytoma cell line. Brain Res 222: 225–233

Hatanaka H, Arimatsu Y (1984) Monoclonal antibodies to tyrosine hydroxylase from rat pheochromocytoma PC12h cells with special reference to nerve growth factor-mediated increase of the immunoprecipitable enzymes. Neurosci Res 1: 253–263

Hatanaka H, Otten U, Thoenen H (1978) Nerve growth factor-mediated selective induction of ornithine decarboxylase in rat pheochromocytoma; a cyclic AMP-independent process. FEBS Lett 92: 313–316

Hayes TE, Kitchen AM, Cochran BH (1987) Inducible binding of a factor to the c-*fos* regulatory region. Proc Natl Acad Sci USA 84: 1272–1276

Holt JT, Gopal TV, Moulton AD, Nienhuis AW (1986) Inducible production of c-*fos* antisense RNA inhibits 3T3 cell proliferation. Proc Natl Acad Sci USA 83: 4794–4798

Huff KR, Guroff G (1979) Nerve growth factor-induced reduction of epidermal growth factor responsiveness and epidermal growth factor receptors in PC12 cells: an aspect of cell differentiation. Biochem Biophys Res Commun 89: 175–180

Improta T, Salvatore AM, Di Luzio A, Romeo G, Coccia EM, Calissano P (1988) IFN-gamma facilitates NGF-induced neuronal differentiation in PC12 cells. Exp Cell Res 179: 1–9

Inoue N, Hatanaka H (1982) Nerve growth factor induces specific enkephalin binding sites in a nerve cell line. J Biol Chem 257: 9238–9241

Inoue N, Matsui H, Hatanaka H (1988) Nerve growth factor induces Na^+, K^+ – ATPase in a nerve cell line. J Neurochem 50: 230–236

Jacobs JR, Stevens JK (1986) Changes in the organization of the neuritic cytoskeleton during nerve growth factor-activated differentiation of PC12 cells: a serial electron microscopic study of the development and control of neurite shape. J Cell Biol 103: 895–906

Jessell TM (1988) Adhesion molecules and the hierarchy of neural development. Neuron 1: 3–13

Johnson EM Jr, Taniuchi M, Clark HB, Springer JE, Koh S, Tayrien MW, Loy R (1987) Demonstration of the retrograde transport of nerve growth factor receptor in the peripheral and central nervous system. J Neurosci 7: 923–929

Jumblatt JE, Tischler AS (1982) Regulation of muscarinic ligand binding sites by nerve growth factor in PC12 phaeochromocytoma cells. Nature 297: 152–154

Kalman D, Wong B, Horvai AE, Cline MJ, O'Lague PH (1990) Nerve growth factor acts through cAMP-dependent protein kinase to increase the number of sodium channels in PC12 cells. Neuron (in press)

Kerr LD, Holt JT, Matrisian LM (1988) Growth factors regulate transin gene expression by c-*fos*-dependent and c-*fos*-independent pathways. Science 242: 1424–1427

Koizumi S, Contreras ML, Matsuda Y, Hama T, Lazarovici P, Guroff G (1988) K-252a: a specific inhibitor of the action of nerve growth factor on PC 12 cells. J Neurosci 8: 715–721

Kruijer W, Schubert D, Verma IM (1985) Induction of the proto-oncogene *fos* by nerve growth factor. Proc Natl Acad Sci USA 82: 7330–7334

Kujubu DA, Lim RW, Varnum BC, Herschman HR (1987) Induction of transiently expressed genes in PC-12 pheochromocytoma cells. Oncogene 1: 257–262

Lacal JC, Moscat J, Aaronson SA (1987) Novel source of 1,2-diacylglycerol elevated in cells transformed Ha-*ras* oncogene. Nature 330: 269–272

Lagenaur C, Lemmon V (1987) An L1-like molecule, the 8D9 antigen, is a potent substrate for neurite extension. Proc Natl Acad Sci USA 84: 7753–7757

Lakshmanan J (1978) Is there a second messenger for nerve growth factor-induced phosphatidylinositol turnover in rat superior cervical ganglia. Brain Res 157: 173–177

Lakshmanan J (1979) Post-synaptic PI-effect of nerve growth factor in rat superior cervical ganglia. J Neurochem 32: 1599–1601

Lander AD, Tomaselli K, Calof AL, Reichardt LF (1983) Studies on extracellular matrix components that promote neurite outgrowth. Cold Spring Harbor Symp Quant Biol 48: 611–623

Landreth GE, Rieser GD (1985) Nerve growth factor- and epidermal growth factor-stimulated phosphorylation of a PC12 cytoskeletally associated protein in situ. J Cell Biol 100: 677–683

Landreth GE, Williams LK (1987) Nerve growth factor stimulates the phosphorylation of a 250 kDa cytockeletal protein in cell-free extracts of PC12 cells. Neurochem Res 12: 943–950

Lau LF, Nathans D (1985) Identification of a set of genes expressed during the G0/G1 transition of cultured mouse cells. EMBO J 4: 3145–3151

Lazarovici P, Dickens G, Kuzuya H, Guroff G (1987) Long-term, heterologous down-regulation of the epidermal growth factor receptor in PC12 cells by nerve growth factor. J Cell Biol 104: 1611–1621

Lazarovici P, Levi BZ, Lelkes PI, Koizumi S, Fujita K, Matsuda Y, Ozato K, Guroff G (1989) K-252a inhibits the increase in c-*fos* transcription and the increase in intracellular calcium produced by nerve growth factor in PC12 cells. J Neurosci Res 23: 1–8

Lee KY, Seeley PJ, Müller TH, Helmer-Matyjek E, Sabban E, Goldstein M, Greene LA (1985) Regulation of tyrosine hydroxylase phosphorylation in PC12 pheochromocytoma cells by elevated K^+ and nerve growth factor. Evidence for different mechanisms of action. Mol Pharmacol 28: 220–228

Lee V, Trojanowski JQ, Schlaepfer WW (1982) Induction of neurofilament triplet proteins in PC12 cells by nerve growth factor. Brain Res 238: 169–180

Lee VM (1985) Neurofilament protein abnormalities in PC12 cells: comparison with neurofilament proteins of normal cultured rat sympathetic neurons. J Neurosci 5: 3039–3046

Lee VM, Page C (1984) The dynamics of nerve growth factor-induced neurofilament and vimentin filament expression and organization in PC12 cells. J Neurosci 4: 1705–1714

Leonard DG, Ziff EB, Greene LA (1987) Identification and characterization of mRNAs regulated by nerve growth factor in PC12 cells. Mol Cell Biol 7: 3156–3167

Leonard DG, Gorham JD, Cole P, Greene LA, Ziff EB (1988) A nerve growth factor-regulated messenger RNA encodes a new intermediate filament protein. J Cell Biol 106: 181–193

Letourneau PC (1979) Cell-substratum adhesion of neurite growth cones, and its role in neurite elongation. Exp Cell Res 124: 127–138

Letourneau PC (1981) Immunocytochemical evidence for colocalization in neurite growth cones of actin and myosin and their relationship to cell—substratum adhesions. Dev Biol 85: 113–122

Levi A, Shechter Y, Neufeld EJ, Schlessinger J (1980) Mobility, clustering, and transport of nerve growth factor in embryonal sensory cells and in a sympathetic neuronal cell line. Proc Natl Acad Sci USA 77: 3469–3473

Levi A, Eldridge JD, Paterson BM (1985) Molecular cloning of a gene sequence regulated by nerve growth factor. Science 229: 393–395

Lewis SA, Sherline P, Cowan NJ (1986) A cloned cDNA encoding MAP1 detects a single copy gene in mouse and a brain-abundant RNA whose level decreases during development. J Cell Biol 102: 2106–2114

Lim RW, Varnum BC, Herschman HR (1987) Cloning of tetradecanoyl phorbol ester-induced 'primary response' sequences and their expression il density-arrested Swiss 3T3 cells and a TPA non-proliferative variant. Oncogene 1: 263–270

Lindenbaum MH, Carbonetto S, Mushynski WE (1987) Nerve growth factor enhances the synthesis, phosphorylation, and metabolic stability of neurofilament proteins in PC12 cells. J Biol Chem 262: 605–610

Lindenbaum MH, Carbonetto S, Grosveld F, Flavell D, Mushynski WE (1988) Transcriptional and post-transcriptional effects of nerve growth factor on expression of the three neurofilament subunits in PC-12 cells. J Biol Chem 263: 5662–5667

Lucas CA, Czlonkowska A, Kreutzberg GW (1980) Regulation of acetylcholinesterase by nerve growth factor in the pheochromocytoma PC12 cell line. Neurosci Lett 18: 333–337

Machida CM, Rodland KD, Matrisian L, Magun BE, Ciment G (1989) NGF induction of the gene encoding the protease transin accompanies neuronal differentiation in PC12 cells. Neuron 2: 1587–1596

Maher PA (1988) Nerve growth factor induces protein-tyrosine phosphorylation. Proc Natl Acad Sci USA 85: 6788–6791

Maher PA (1989) Role of protein tyrosine phosphorylation in the NGF response. J Neurosci Res 24: 29–37

Makanthappa NK, Patterson PH (1989) Thy-1 dimerization, Thy-1-derived peptides, and neurite outgrowth. Soc Neurosci Abstr 15: 651 (abstract)

Mandel G, Cooperman SS, Maue RA, Goodman RH, Brehm P (1988) Selective induction of brain type II Na + channels by nerve growth factor. Proc Natl Acad Sci USA 85: 924–928

Margolis RK, Salton SR, Margolis RU (1987) Effects of nerve growth factor-induced differentiation on the heparan sulfate of PC12 pheochromocytoma cells and comparison with developing brain. Arch Biochem Biophys 257: 107–114

Marsh L, Letourneau PC (1984) Growth of neurites without filopodial or lamellipodial activity in the presence of cytochalasin B.J Cell Biol 99: 2041–2047

Masiakowski P, Shooter EM (1988) Nerve growth factor induces the genes for two proteins related to a family of calcium-binding proteins in PC12 cells. Proc Natl Acad Sci USA 85: 1277–1281

Matsuda Y, Guroff G (1987) Purification and mechanism of activation of a nerve growth factor-sensitive S6 kinase from PC12 cells. J Biol Chem 262: 2832–2844

Matsuda Y, Nakanishi N, Dickens G, Guroff G (1986) A nerve growth factor-sensitive S6 kinase in cell-free extracts from PC12 cells. J Neurochem 47: 1728–1734

Max SR, Schwab M, Dumas M, Thoenen H (1978) Retrograde axonal transport of nerve growth factor in the ciliary ganglion of the chick and the rat. Brain Res 159: 411–415

McGuire JC, Greene LA (1980) Stimulation by nerve growth factor of specific protein synthesis in rat PC12 pheochromocytoma cells. Neuroscience 5: 179–189

McGuire JC, Greene LA, Furano AV (1978) NGF stimulates incorporation of fucose or glucosamine into an external glycoprotein in cultured rat PC12 pheochromocytoma cells. Cell 15: 357–365

McTigue M, Cremins J, Halegoua S (1985) Nerve growth factor and other agents mediate phosphorylation and activation of tyrosine hydroxylase. A convergence of multiple kinase activities. J Biol Chem 260: 9047–9056

Milbrandt J (1987) A nerve growth factor-induced gene encodes a possible transcriptional regulatory factor. Science 238: 797–799

Milbrandt J (1988) Nerve growth factor induces a gene homologous to the glucocorticoid receptor gene. Neuron 1: 183–188

Mitchison T, Kirschner M (1988) Cytoskeletal dynamics and nerve growth. Neuron 1: 761–772

Mitsuka M, Hatanaka H (1984) Increase of carbamylcholine-induced $^{22}Na^{+}$ influx into pheochromocytoma PC12h cells by nerve growth factor. Brain Res 314: 255–260

Montminy MR, Sevarino KA, Wagner JA, Mandel G, Goodman RH (1986) Identification of a cyclic-AMP-responsive element within the rat somatostatin gene. Proc Natl Acad Sci USA 83: 6682–6686

Morgan JI, Curran T (1986) Role of ion flux in the control of c-*fos* expression. Nature 322: 552–555

Morgan JI, Curran T (1989) Stimulus-transcription coupling in neurons: role of cellular immediate-early genes. TINS 12: 459–462

Müller R, Bravo R, Burckhardt J, Curran T (1984) Induction of c-*fos* gene and protein by growth factors precedes activation of c-*myc*. Nature 312: 716–720

Nakanishi N, Guroff G (1985) Nerve growth factor-induced inccrease in the cell-free phosphorylation of a nuclear protein in PC12 cells. J Biol Chem 260: 7791–7799

Naoi M, Shibahara K, Suzuki H, Nagatsu T (1988) Specific binding of glycosylated beta-galactosidase to rat clonal pheochromocytoma PC12h cells: effects of nerve growth factor and gangliosides. J Neurochem 50: 1297–1301

Narumi S, Fujita T (1978) Stimulatory effects of substance P and nerve growth factor (NGF) on neurite outgrowth in embryonic chick dorsal root ganglia. Neuropharmacology 17: 73–76

Neer EJ, Clapham DE (1988) Roles of G protein subunits in transmembrane signalling. Nature 333: 129–134

Nikodijevic B, Nikodijevic O, Yu MY, Pollard H, Guroff G (1975) The effect of nerve growth factor on cycle AMP levels in superior cerival ganglia of the rat. Proc Natl Acad Sci USA 72: 4769–4771

Nishikura K, Murray JM (1987) Antisense RNA of proto-oncogene c-*fos* blocks renewed growth of quiescent 3T3 cells. Mol Cell Biol 7: 639–649

Noda M, Ko M, Ogura A, Liu DG, Amano T, Takano T, Ikawa Y (1985) Sarcoma viruses carrying *ras* oncogenes induce differentiation-associated properties in a neuronal cell line. Nature 318: 73–75

O'Lague PH, Huttner SL (1980) Physiological and morphological studies of rat pheochromocytoma cells (PC12) chemically fused and grown in culture. Proc Natl Acad Sci USA 77: 1701–1705

Okabe S, Hirokawa N (1988) Microtubule dynamics in nerve cells: analysis using microinjection of biotinylated tubulin into PC12 cells. J Cell Biol 107: 651–664

Pandiella-Alonso A, Malgaroli A, Vicentini LM, Meldolesi J (1986) Early rise of cytosolic Ca^{2+} induced by NGF in PC12 and chromaffin cells. FEBS Lett 208: 48–51

Parysek LM, Goldman RD (1987) Characterization of intermediate filaments in PC12 cells. J Neurosci 7: 781–791

Pfenninger KH, Johnson MP (1981) Nerve growth factor stimulates phospholipid methylation in growing neurites. Proc Natl Acad Sci USA 78: 7797–7800

Portier MM, Croizat B, Gros F (1982) A sequence of changes in cytoskeletal components during neuroblastoma differentiation. FEBS Lett 146: 283–288

Portier MM, Brachet P, Croizat B, Gros F (1983) Regulation of peripherin in mouse neuroblastoma and rat PC12 pheochromocytoma cell lines. Devi Neurosci 6: 215–226

Possenti R, Eldridge JD, Paterson BM, Grasso A, Levi A (1989) A protein induced by NGF in PC12 cells is stored in secretory vesicles and released through the regulated pathway. EMBO J 8: 2217–2223

Prentice HM, Moore SE, Dickson JG, Doherty P, Walsh FS (1987) Nerve growth factor-induced changes in neural cell adhesion molecule (N-CAM) in PC12 cells. EMBO J 6: 1859–1863

Race HM, Wagner JA (1985) Nerve growth factor affects cyclic AMP metabolism, but not by directly stimulating adenylate cyclase activity. J Neurochem 44: 1588–1592

Rausch DM, Dickens G, Doll S, Fujita K, Koizumi S, Rudkin BB, Tocco M, Eiden LE, Guroff G (1989) Differentiation of PC12 cells with v-*src*: comparison with nerve growth factor. J Neurosci Res 24: 49–58

Reinhold DS, Neet KE (1989) The lack of a role for protein kinase C in neurite extension and in the induction of ornithine decarboxylase by nerve growth factor in PC12 cells. J Biol Chem 264: 3538–3544

Riabowol KT, Vosatka RJ, Ziff EB, Lamb NJ, Feramisco JR (1988) Microinjection of fos-specific antibodies blocks DNA synthesis in fibroblast cells. Mol Cell Biol 8: 1670–1676

Richter-Landsberg C, Jastorff B (1986) The role of cAMP in nerve growth factor-promoted neurite outgrowth in PC12 cells. J Cell Biol 102: 821–829

Richter-Landsberg C, Greene LA, Shelanski ML (1985) Cell surface Thy-1-cross-reactive glycoprotein in cultured PC12 cells: modulation by nerve growth factor and association with the cytoskeleton. J Neurosci 5: 468–476

Rieger F, Shelanski ML, Greene LA (1980) The effects of nerve growth factor on acetylcholinesterase and its multiple forms in cultures of rat PC12 pheochromocytoma cells: increased total specific activity and appearance of the 16S molecular form. Dev Biol 76: 238–243

Romano C, Nichols RA, Greengard P (1987) Synapsin I in PC12 cells. II. Evidence for regulation by NGF of phosphorylation at a novel site. J Neurosci 7: 1300–1306

Rowland EA, Müller TH, Goldstein M, Greene LA (1987) Cell-free detection and characterization of a novel nerve growth factor-activated protein kinase in PC12 cells. J Biol Chem 262: 7504–7513

Rudy B, Kirschenbaum B, Rukenstein A, Greene LA (1987) Nerve growth factor increases the number of functional Na channels and induces TTX-resistant Na channels in PC12 pheochromocytoma cells. J Neurosci 7: 1613–1625

Rutishauser U, Jessell TM (1988) Cell adhesion molecules in vertebrate neural development. Physiol Rev 68: 819–857

Rydel RE, Greene LA (1987) Acidic and basic fibroblast growth factors promote stable neurite outgrowth and neuronal differentiation in cultures of PC12 cells. J Neurosci 7: 3639–3653

Saarma M, Toots U, Raukas E, Zhelkovsky A, Pivazian A, Neuman T (1986) Nerve growth factor induces changes in (2′–5′)oligo(A) synthetase and 2′-phosphodiesterase activities during differentiation of PC12 pheochromocytoma cells. Exp Cell Res 166: 229–236

Sajovic P, Ennulat DJ, Shelanski ML, Greene LA (1987) Isolation of NILE glycoprotein-related cDNA probes. J Neurochem 49: 756–763

Saltiel AR, Cuatrecasas P (1988) In search of a second messenger for insulin. Am J Physiol 255: C1–C11

Sano K, Kikuchi A, Matsui Y, Teranishi Y, Takai Y (1989) Tissue-specific expression of a novel GTP-binding protein (smg p25A) mRNA and its increase by nerve growth factor and cyclic AMP in rat pheochromocytoma PC-12 cells. Biochem Biophys Res Commun 158: 377–385

Sassone-Corsi P, Visvader J, Ferland L, Mellon PL, Verma IM (1988) Induction of proto-oncogene *fos* transcription through the adenylate cyclase pathway: characterization of a cAMP-responsive element. Genes Dev 2: 1529–1538

Sassone-Corsi P, Der CJ, Verma IM (1989) *ras*-Induced neuronal differentiation of PC12 cells: possible involvement of *fos* and *jun*. Mol Cell Biol 9: 3174–3183

Schonthal A, Herrlich P, Rahmsdorf HJ, Ponta H (1988) Requirement for *fos* gene expression in the transcriptional activation of collagenase by other oncogenes and phorbol esters. Cell 54: 325–334

Schubert D, Whitlock C (1977) Alteration of cellular adhesion by nerve growth factor. Proc Natl Acad Sci USA 74: 4055–4058

Schubert D, Heinemann S, Kidokoro Y (1977) Cholinergic metabolism and synapse formation by a rat nerve cell line. Proc Natl Acad Sci USA 74: 2579–2583

Schubert D, LaCorbiere M, Whitlock C, Stallcup W (1978) Alterations in the surface properties of cells responsive to nerve growth factor. Nature 273: 718–723

Schubert D, Jin L-W, Saitoh T, Cole G (1989) The regulation of amyloid β protein precursor secretion and its modulatory role in cell adhesion. Neuron 3: 689–694

Schwarz MA, Brown PJ, Eveleth DD, Bradshaw RA (1989) Modulation of growth factor induced fiber outgrowth in rat pheochromocytoma (PC12) cells by a fibronectin receptor antibody. J Cell Physiol 138: 121–128

Seeley PJ, Rukenstein A, Connolly JL, Greene LA (1984) Differential inhibition of nerve growth factor and epidermal growth factor effects on the PC12 pheochromocytoma line. J Cell Biol 98: 417–426

Sheng M, Dougan ST, McFadden G, Greenberg ME (1988) Calcium and growth factor pathways of c-*fos* transcriptional activation require distinct upstream regulatory sequences. Mol Cell Biol 8: 2787–2796

Shibanuma M, Kuroki T, Nose K (1987) Inhibition of proto-oncogene c-*fos* transcription by inhibitors of protein kinase C and ion transport. Eur J Biochem 164: 15–19

Skaper SD, Varon S (1984) Nerve growth factor stimulates phospholipid methylation in target ganglionic neurons independently of the cyclic AMP and sodium pump responses. J Neurochem 42: 116–122

Skaper SD, Bottenstein JE, Varon S (1979) Effects of nerve growth factor on cyclic AMP levels in embryonic chick dorsal root ganglia following factor deprivation. J Neurochem 32: 1845–1851

Skene JH (1989) Axonal growth-associated proteins. Annu Rev Neurosci 12: 127–156

Smith SJ (1988) Neuronal cytomechanics: the actin-based motility of growth cones. Science 242: 708–715

Smith MR, DeGudicibus SJ, Stacey DW (1986) Requirement for c-*ras* proteins during viral oncogene transformation. Nature 320: 540–543

Smith DS, King CS, Pearson E, Gittinger CK, Landreth GE (1989) Selective inhibition of nerve growth factor-stimulated protein kinases by K-252a and 5′-S-methyladenosine in PC12 cells. J Neurochem 53: 800–806

Sobue K, Kanda K (1988) Localization of pp60^{csrc} in growth cone of PC12 cell. Biochem Biophys Res Commun 157: 1383–1389

Sonnenberg JL, Rauscher FJ III, Morgan JI, Curran T (1989) Regulation of proenkephalin by *fos* and *jun*. Science 246: 1622–1625

Stacey DW, Watson T, Kung HF, Curran T (1987) Microinjection of transforming *ras* protein induces c-*fos* expression. Mol Cell Biol 7: 523–527

Stallcup WB, Beasley L (1985) Involvement of the nerve growth factor-inducible large external glycoprotein (NILE) in neurite fasciculation in primary cultures of rat brain. Proc Natl Acad Sci USA 82: 1276–1280

Stein R, Orit S, Anderson DJ (1988) The induction of a neural-specific gene, *SCG10*, by nerve growth factor in PC12 cells is transcriptional, protein synthesis dependent, and glucocorticoid inhibitable. Dev Biol 127: 316–325

Strassman RJ, Letourneau PC, Wessells NK (1973) Elongation of axons in an agar matrix that does not support cell locomotion. Exp Cell Res 81: 482–487

Sugimoto Y, Noda M, Kitayama H, Ikawa Y (1988) Possible involvement of two signaling pathways in induction of neuron-associated properties by v-Ha-*ras* gene in PC12 cells. J Biol Chem 263: 12102–12108

Takahashi M, Tsukui H, Hatanaka H (1985) Neuronal differentiation of Ca^{2+} channel by nerve growth factor. Brain Res 341: 381–384

Taniguchi T, Morisawa K, Ogawa M, Yamamoto H, Fujimoto S (1988) Decrease in the level of poly(ADP-ribose) synthetase during nerve growth factor-promoted neurite outgrowth in rat pheochromocytoma PC12 cells. Biochem Biophys Res Commun 154: 1034–1040

Thoenen H, Angeletti PU, Levi-Montalcini R, Kettler R (1971) Selective induction by nerve growth factor of tyrosine hydroxylase and dopamine-hydroxylase in the rat superior cervical ganglia. Proc Natl Acad Sci USA 68: 1598–1602

Thompson MA, Ziff EB (1989) Structure of the gene encoding peripherin, an NGF-regulated neuronal-specific type III intermediate filament protein. Neuron 2: 1043–1053

Tiercy JM, Shooter EM (1986) Early changes in the synthesis of nuclear and cytoplasmic proteins are induced by nerve growth factor in differentiating rate PC12 cells. J Cell Biol 103: 2367–2378

Tirone F, Shooter EM (1989) Early gene regulation by nerve growth factor in PC12 cells: induction of an interferon-related gene. Proc Natl Acad Sci USA 86: 2088–2092

Tischler AS, Lee YC, Slayton VW, Bloom SR (1982) Content and release of neurotensin in PC12 pheochromocytoma cell cultures: modulation by dexamethasone and nerve growth factor. Regul Pept 3: 415–421

Toda T, Uno I, Ishikawa T, Powers S, Kataoka T, Broek D, Cameron S, Broach J, Matsumoto K, Wigler M (1985) In yeast, *RAS* proteins are controlling elements of adenylate cyclase. Cell 40: 27–36

Togari A, Guroff G (1985) Partial purification and characterization of a nerve growth factor-sensitive kinase and its substrate from PC12 cells. J Biol Chem 260: 3804–3811

Togari A, Dickens G, Kuzuya H, Guroff G (1985) The effect of fibroblast growth factor on PC12 cells. J Neurosci 5: 307–316

Traynor AE (1984) The relationship between neurite extension and phospholipid metabolism in PC12 cells. Brain Res 316: 205–210

Traynor AE, Schubert D, Allen WR (1982) Alterations of lipid metabolism in response to nerve growth factor. J Neurochem 39: 1677–1683

Treisman R (1985) Transient accumulation of c-*fos* RNA following serum stimulation requires a conserved 5′ element and c-*fos* 3′ sequences. Cell 42: 889–902

Turner DC, Flier LA, Carbonetto S (1989) Identification of a cell-surface protein involved in PC12 cell-substratum adhesion and neurite outgrowth on laminin and collagen. J Neurosci 9: 3287–3296

Van Hooff CO, De Graan PN, Oestreicher AB, Gispen WH (1988) B-50 phosphorylation and polyphosphoinositide metabolism in nerve growth cone membranes. J Neurosci 8: 1789–1795

Van Hooff CO, Holthuis JC, Oestreicher AB, Boonstra J, De Graan PN, Gispen WH (1989) Nerve growth factor-induced changes in the intracellular localization of the protein kinase C substrate B-50 in pheochromocytoma PC12 cells. J Cell Biol 108: 1115–1125

Vinores SA, Marangos PJ, Parma AM, Guroff G (1981) Increased levels of neuron-specific enolase in PC12 pheochromocytoma cells as a result of nerve growth factor treatment. J Neurochem 37: 597–600

Visvader J, Sassone-Corsi P, Verma IM (1988) Two adjacent promoter elements mediate nerve growth factor activation of the c-*fos* gene and bind distinct nuclear complexes. Proc Natl Acad Sci USA 85: 9474–9478

Volonte C, Rukenstein A, Loeb DM, Greene LA (1989) Differential inhibition of nerve growth factor responses by purine analogues: correlation with inhibition of a nerve growth factor-activated protein kinase. J Cell Biol 109: 2395–2403

Wu BY, Fodor EJ, Edwards RH, Rutter WJ (1989) Nerve growth factor induces the proto-oncogene c-*jun* in PC12 cells. J Biol Chem 264: 9000–9003

Yamamoto KR, Alberts BM (1976) Steroid receptors: elements for modulation of eukaryotic transcription. Annu Rev Biochem 45: 721–746

Yankner BA, Shooter EM (1979) Nerve growth factor in the nucleus: interaction with receptors on the nuclear membrane. Proc Natl Acad Sci USA 76: 1269–1273

Young S, Parker PJ, Ullrich A, Stabel S (1987) Down-regulation of protein kinase C is due to an increased rate of degradation. Biochem J 244: 775–779

Yu CL, Tsai MH, Stacey DW (1988) Cellular *ras* activity and phospholipid metabolism. Cell 52: 63–71

Yu MW, Tolson NW, Guroff G (1980) Increased phosphorylation of specific nuclear proteins in superior cervical ganglia and PC12 cells in response to nerve growth factor. J Biol Chem 255: 10481–10492

Zigmond RE, Schwarzschild MA, Rittenhouse AR (1989) Acute regulation of tyrosine hydroxylase by nerve activity and by neurotransmitters via phosphorylation. Annu Rev Neurosci 12: 415–461

Zuber MX, Goodman DW, Karns LR, Fishman MC (1989) The neuronal growth-associated protein GAP-43 induces filopodia in non-neuronal cells. Science 244: 1193–1195

Subject Index

Current Topics in Microbiology and Immunology

Volumes published since 1984 (and still available)

Vol. 112: **Vogt, Peter K.; Koprowski, Hilary (Ed.):** Retroviruses 3. 1984. 19 figs. VII, 115 pp. ISBN 3-540-13307-0

Vol. 113: **Potter, Michael; Melchers, Fritz; Weigert, Martin (Ed.):** Oncogenes in B-Cell Neoplasia. Workshop at the National Cancer Institute, National Institutes of Health, Bethesda, MD March 5–7, 1984. 1984. 65 figs. XIII, 268 pp. ISBN 3-540-13597-9

Vol. 115: **Vogt, Peter K. (Ed.):** Human T-Cell Leukemia Virus. 1985. 74 figs. IX, 266 pp. ISBN 3-540-13963-X

Vol. 116: **Willis, Dawn B. (Ed.):** Iridoviridae. 1985. 65 figs. X, 173 pp. ISBN 3-540-15172-9

Vol. 122: **Potter, Michael (Ed.):** The BALB/c Mouse. Genetics and Immunology. 1985. 85 figs. XVI, 254 pp. ISBN 3-540-15834-0

Vol. 124: **Briles, David E. (Ed.):** Genetic Control of the Susceptibility to Bacterial Infection. 1986. 19 figs. XII, 175 pp. ISBN 3-540-16238-0

Vol. 125: **Wu, Henry C.; Tai, Phang C. (Ed.):** Protein Secretion and Export in Bacteria. 1986. 34 figs. X, 211 pp. ISBN 3-540-16593-2

Vol. 126: **Fleischer, Bernhard; Reimann, Jörg; Wagner, Hermann (Ed.):** Specificity and Function of Clonally Developing T-Cells. 1986. 60 figs. XV, 316 pp. ISBN 3-540-16501-0

Vol. 127: **Potter, Michael; Nadeau, Joseph H.; Cancro, Michael P. (Ed.):** The Wild Mouse in Immunology. 1986. 119 figs. XVI, 395 pp. ISBN 3-540-16657-2

Vol. 128: 1986. 12 figs. VII, 122 pp. ISBN 3-540-16621-1

Vol. 129: 1986. 43 figs., VII, 215 pp. ISBN 3-540-16834-6

Vol. 130: **Koprowski, Hilary; Melchers, Fritz (Ed.):** Peptides as Immunogens. 1986. 21 figs. X, 86 pp. ISBN 3-540-16892-3

Vol. 131: **Doerfler, Walter; Böhm, Petra (Ed.):** The Molecular Biology of Baculoviruses. 1986. 44 figs. VIII, 169 pp. ISBN 3-540-17073-1

Vol. 132: **Melchers, Fritz; Potter, Michael (Ed.):** Mechanisms in B-Cell Neoplasia. Workshop at the National Cancer Institute, National Institutes of Health, Bethesda, MD, USA, March 24–26, 1986. 1986. 156 figs. XII, 374 pp. ISBN 3-540-17048-0

Vol. 133: **Oldstone, Michael B. (Ed.):** Arenaviruses. Genes, Proteins, and Expression. 1987. 39 figs. VII, 116 pp. ISBN 3-540-17246-7

Vol. 134: **Oldstone, Michael B. (Ed.):** Arenaviruses. Biology and Immunotherapy. 1987. 33 figs. VII, 242 pp. ISBN 3-540-17322-6

Vol. 135: **Paige, Christopher J.; Gisler, Roland H. (Ed.):** Differentiation of B Lymphocytes. 1987. 25 figs. IX, 150 pp. ISBN 3-540-17470-2

Vol. 136: **Hobom, Gerd; Rott, Rudolf (Ed.):** The Molecular Biology of Bacterial Virus Systems. 1988. 20 figs. VII, 90 pp. ISBN 3-540-18513-5

Vol. 137: **Mock, Beverly; Potter, Michael (Ed.):** Genetics of Immunological Diseases. 1988. 88 figs. XI, 335 pp. ISBN 3-540-19253-0

Vol. 138: **Goebel, Werner (Ed.):** Intracellular Bacteria. 1988. 18 figs. IX, 179 pp. ISBN 3-540-50001-4

Vol. 139: **Clarke, Adrienne E.; Wilson, Ian A. (Ed.):** Carbohydrate-Protein Interaction. 1988. 35 figs. IX, 152 pp. ISBN 3-540-19378-2

Vol. 140: **Podack, Eckhard R. (Ed.):** Cytotoxic Effector Mechanisms. 1989. 24 figs. VIII, 126 pp. ISBN 3-540-50057-X

Vol. 141: **Potter, Michael; Melchers, Fritz (Ed.):** Mechanisms in B-Cell Neoplasia 1988. Workshop at the National Cancer

Institute, National Institutes of Health, Bethesda, MD, USA, March 23–25, 1988. 1988. 122 figs. XIV, 340 pp.
ISBN 3-540-50212-2

Vol. 142: **Schüpbach, Jörg:** Human Retrovirology. Facts and Concepts. 1989. 24 figs. 115 pp. ISBN 3-540-50455-9

Vol. 143: **Haase, Ashley T.; Oldstone, Michael B. A. (Ed.):** In Situ Hybridization. 1989. 33 figs. XII, 90 pp.
ISBN 3-540-50761-2

Vol. 144: **Knippers, Rolf; Levine, A. J. (Ed.):** Transforming Proteins of DNA Tumor Viruses. 1989. 85 figs. XIV, 300 pp.
ISBN 3-540-50909-7

Vol. 145: **Oldstone, Michael B. A. (Ed.):** Molecular Mimicry. Cross-Reactivity between Microbes and Host Proteins as a Cause of Autoimmunity. 1989. 28 figs. VII, 141 pp. ISBN 3-540-50929-1

Vol. 146: **Mestecky, Jiri; McGhee, Jerry (Ed.):** New Strategies for Oral Immunization. International Symposium at the University of Alabama at Birmingham and Molecular Engineering Associates, Inc. Birmingham, AL, USA, March 21–22, 1988. 1989. 22 figs. IX, 237 pp. ISBN 3-540-50841-4

Vol. 147: **Vogt, Peter K. (Ed.):** Oncogenes. Selected Reviews. 1989. 8 figs. VII, 172 pp.
ISBN 3-540-51050-8

Vol. 148: **Vogt, Peter K. (Ed.):** Oncogenes and Retroviruses. Selected Reviews. 1989. XII, 134 pp. ISBN 3-540-51051-6

Vol. 149: **Shen-Ong, Grace L. C.; Potter, Michael; Copeland, Neal G. (Ed.):** Mechanisms in Myeloid Tumorigenesis. Workshop at the National Cancer Institute, National Institutes of Health, Bethesda, MD, USA, March 22, 1988. 1989. 42 figs. X, 172 pp.
ISBN 3-540-50968-2

Vol. 150: **Jann, Klaus; Jann, Barbara (Ed.):** Bacterial Capsules. 1989. 33 figs. XII, 176 pp.
ISBN 3-540-51049-4

Vol. 151: **Jann, Klaus; Jann, Barbara (Ed.):** Bacterial Adhesins. 1990. 23 figs. XII, 192 pp. ISBN 3-540-51052-4

Vol. 152: **Bosma, Melvin J.; Phillips, Robert A.; Schuler, Walter (Ed.):** The Scid Mouse. Characterization and Potential Uses. EMBO Workshop held at the Basel Institute for Immunology, Basel, Switzerland, February 20–22, 1989. 1989. 72 figs. XII, 263 pp. ISBN 3-540-51512-7

Vol. 153: **Lambris, John D. (Ed.):** The Third Component of Complement. Chemistry and Biology. 1989. 38 figs. X, 251 pp. ISBN 3-540-51513-5

Vol. 154: **McDougall, James K. (Ed.):** Cytomegaloviruses. 1990. 58 figs. IX, 286 pp. ISBN 3-540-51514-3

Vol. 155: **Kaufmann, Stefan H. E. (Ed.):** T-Cell Paradigms in Parasitic and Bacterial Infections. 1990. 24 figs. IX, 162 pp.
ISBN 3-540-51515-1

Vol. 156: **Dyrberg, Thomas (Ed.):** The Role of Viruses and the Immune System in Diabetes Mellitus. 1990. 15 figs. XI, 142 pp.
ISBN 3-540-51918-1

Vol. 157: **Swanstrom, Ronald; Vogt, Peter K. (Ed.):** Retroviruses. Strategies of Replication. 1990. 40 figs. XII, 260 pp.
ISBN 3-540-51895-9

Vol. 158: **Muzyczka, Nicholas (Ed.):** Viral Expression Vectors. 1990. approx. 20 figs. approx. XII, 190 pp.
ISBN 3-540-52431-2

Vol. 159: **Gray, David; Sprent, Jonathan (Ed.):** Immunological Memory. 1990. 38 figs. XII, 156 pp. ISBN 3-540-51921-1

Vol. 160: **Oldstone, Michael B. A.; Koprowski, Hilary (Ed.):** Retrovirus Infections of the Nervous System. 1990. 16 figs. XII, 176 pp.
ISBN 3-540-51939-4

Vol. 161: **Racaniello, Vincent R. (Ed.):** Picornaviruses. 1990. 12 figs. X, 194 pp.
ISBN 3-540-52429-0

Vol. 162: **Roy, Polly; Gorman, Barry M. (Ed.):** Bluetongue Viruses. 1990. 37 figs. X, 200 pp. ISBN 3-540-51922-X

Vol. 163: **Moyer, Richard W.; Turner, Peter C. (Ed.):** Poxviruses. 1990. 23 figs. X, 210 pp.
ISBN 3-540-52430-4

Vol. 164: **Bækkeskov, Steinunn; Hansen, Bruno (Ed.):** Human Diabetes. 1990. 9 figs. X, 198 pp. ISBN 3-540-52652-8